The Romance of Salt

OTHER TITLES

Boman Desai	*A Woman Madly in Love*
Frank Simoes	*Frank Unedited*
Frank Simoes	*Frank Simoes' Goa*
Harinder Baweja (ed.)	*Most Wanted: Profiles of Terror*
J.N. Dixit (ed.)	*External Affairs: Cross-Border Relations*
Khushwant Singh	*Death at My Doorstep*
M.J. Akbar	*India: The Siege Within*
M.J. Akbar	*Kashmir: Behind the Vale*
M.J. Akbar	*Nehru: The Making of India*
M.J. Akbar	*Riot after Riot*
M.J. Akbar	*The Shade of Swords*
M.J. Akbar	*By Line*
Meghnad Desai	*Nehru's Hero Dilip Kumar: In the Life of India*
Nayantara Sahgal (ed.)	*Before Freedom: Nehru's Letters to His Sister*
Rohan Gunaratna	*Inside Al Qaeda*
Rifaat Hussain, J.N. Dixit Julie Sirrs, Ajai Shukla Anand Giridharadas Rahimullah Yusufzai John Jennings	*Afghanistan and 9/11*
Eric S. Margolis	*War at the Top of the World*
Maj. Gen. Ian Cardozo	*Param Vir: Our Heroes in Battle*
Mushirul Hasan	*India Partitioned. 2 Vols*
Mushirul Hasan	*John Company to the Republic*
Mushirul Hasan	*Knowledge Power and Politics*
Prafulla Roy, tr. John W. Hood	*In the Shadow of the Sun*
Rachel Dwyer	*Yash Chopra: Fifty Years of Indian Cinema*
Saad Bin Jung	*Wild Tales from the Wild*
Satish Jacob	*From Hotel Palestine Baghdad*
Sujata S. Sabnis	*A Twist in Destiny*
V.N. Rai	*Curfew in the City*
Vaibhav Purandare	*Sachin Tendulkar: A Difinitive Biography*

FORTHCOMING TITLES:

Duff Hart-Davis	*Honorary Tiger: The Life of Billy Arjan Singh*
Rashme Sehgal	*The Sentinel*

ANIL DHARKER

LOTUS COLLECTION
ROLI BOOKS

Lotus Collection

© Anil Dharker 2005
All rights reserved. No part of this publication may be reproduced or transmitted, in any form or by any means, without the prior permission of the publisher.

This edition first published in 2005

The Lotus Collection
An imprint of
Roli Books Pvt. Ltd.
M-75, G.K. II Market
New Delhi 110 048
Phones: ++91-11-2921 2271, 2921 2782
2921 0886, Fax: ++91-11-2921 7185
E-mail: roli@vsnl.com; Website: rolibooks.com
Also at
Bangalore, Varanasi, Jaipur and the Netherlands

Extracts from *Young India* issues dated January to May 1930, reproduced with permission of the Navjivan Trust, Ahmedabad.
Extract from Thomas Webber, Pgs. 113-114, reproduced from *On The Salt March*, published by HarperCollins India Pvt. Ltd., 1997.

Cover design: Sneha Pamneja
Layout design: Kumar Raman

ISBN: 81-7436-398-X
Rs 395/-

Typeset in Minion by Roli Books Pvt. Ltd. and printed at Presstech Litho Pvt. Ltd., Noida - 201 301

DEDICATION

For Minal and Rani
You are
Simply the best

CONTENTS

Acknowledgements	ix

Part 1
SALT MARCH TO DANDI — 1

Prologue	3
The Ashramites of Sabarmati	9
The Choice of Weapons	20
The Girding of Loins	28
The Medium and the Message	43
The Final Countdown	54
The March to Dandi	81
Epilogue	108

Part 2
SALT'S MARCH THROUGH HISTORY — 117

The Stuff of Life	119
Salt Through History	129
Salt Wars	139
Dead Sea, Live Seas	148
Superstition and Myth	153
The Way of All Flesh	161
Literature, Art and the Cellars Market	172
See How It Runs	196
Ladies of Salt	207
Giving Life to Life	217
Select Bibliography	226

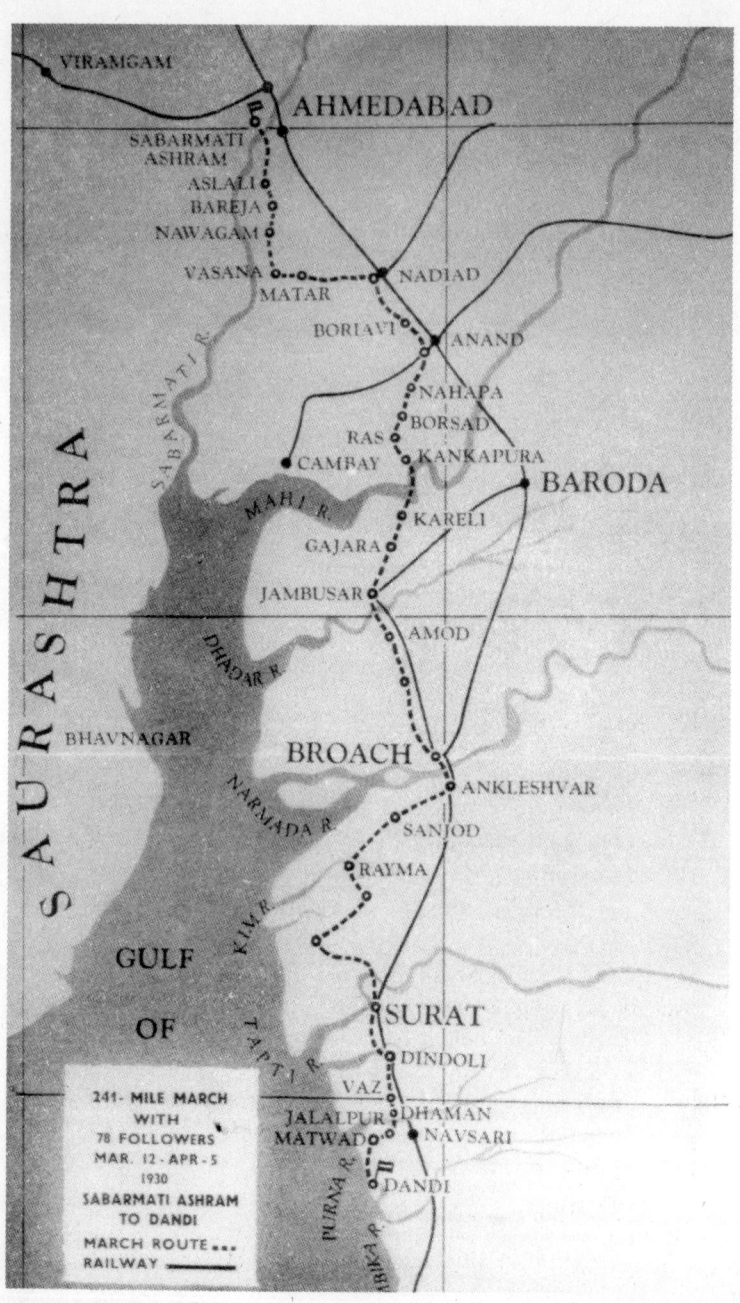

ACKNOWLEDGEMENTS

'Salt? A book on **salt**? Should be about ten pages long,' said a friend when I put the idea to her. Perhaps it was a gender thing. I broached the idea with a male friend. '**Salt?**' he said, 'You aren't being serious, of course.' It was not a gender thing.

Luckily, some people were more positive. 'Hey, that's a great idea,' said Satish Sohoni, who heads Tata Salt, India's biggest producer. His enthusiasm was, if anything, greater than mine. The Managing Director of Tata Chemicals, Prasad Menon, was equally welcoming. Pramod Kapoor, Publisher of Roli Books and Renuka Chatterjee, its Chief Editor, responded as warmly and as quickly. The contract for publishing was signed, sealed and delivered in record time.

The rush came about because one date couldn't be changed. As you will see from the book, salt has many important landmarks through the centuries, but its Indian brush with history is indelibly fixed in our mind with Mahatma Gandhi's march to Dandi. It made eminent sense to launch the book on the March's anniversary.

It made no sense at all in every other way because it gave me just three months from conception to delivery, and that's premature birth in any language. The urgency came about because one particular conversation with R. Gopalakrishnan, energetic tennis partner and valuable friend, took place only around December. It was in the locker room after a game, the place where in the sweat and the grime, great ideas often emerge. Our stream of consciousness probably went like this: Great game of tennis, muggy day though. Muggy ⟶ Sweat ⟶ Salt ⟶ Dandi March ⟶ 75th Anniversary ⟶ Book on Salt. The result you now hold in your hands.

I hope it's worthy of all the help that was given so unstintingly. Tata Chemicals' support was essential to the project and it was offered whole-heartedly, from Satish Sohoni's unflagging interest, to the invaluable inputs from his team at Mithapur, particularly Vivek Talwar, Alka Talwar, Satish Trivedi and Rehana Shiekh. The work they are doing there for the community is a sterling example of the Tata group's sense of corporate social responsibility. Many thanks too to Ismail Momin and Ashok Thakur for help on technical matters, to Minal Dharker for unravelling myths and recipes and Sugatha Menon for very useful legwork. Finally, a great big thanks to Rani Dharker, who neglected her own writing to help with research. Without her, this book wouldn't have happened.

To salt, then, without which we would lead a very tasteless life.

Mumbai   Anil Dharker
March 2005

Part 1

SALT MARCH TO DANDI

PROLOGUE

The Dandi Salt March of 12 March 1930 need not have happened at Dandi, it need not have been about salt and it need not have been a march. What made the event a landmark in India's freedom struggle was the synthesis of these disparate elements into a unit which was so breathtaking in its simplicity that it captured the world's imagination. But then, Mahatma Gandhi always had a gift for striking imagery.

The march's genesis goes back, in a sense, to the appointment of the Simon Commission. On 2 November 1927, prominent Indian political leaders like Gandhi, Motilal Nehru, Dr M.A. Ansari and M.A. Jinnah were summoned by the Viceroy and handed over a document announcing the appointment of a Royal Commission to be headed by Sir John Simon. Gandhi, who had to travel over fifteen hundred kilometres to get to Delhi, wondered why the document couldn't be posted. Especially because the composition of the commission and its terms of reference were so disappointing.

The surprising element in the appointment of the commission was its timing: the Indian Reforms Act of 1919 under which the British governed India, contained a clause for a review of the constitutional position after ten years. The two involved parties saw this bit of legalese in their own light: British politicians of the ruling Conservative party saw it as an escape clause; Indian nationalists saw it as a route to future concessions. Either way, the review wasn't due until 1929. In bringing forward the date by two years, the British Government was being too clever by half: it sought to convey the impression that it was being sympathetic to nationalist aspirations; while in fact, the real reason was different. The portents in Britain clearly indicated that the Labour government would win the forthcoming general elections. The Conservatives wanted to ensure that they rather than Labour appointed the commission for India.

People, of course, saw through the ruse. Anger mounted even further when Lord Birkenhead, Secretary of State for India, announced the composition of the commission. Except for the chairman, Sir John Simon, its other members were seen as 'second flight men'. Worse, all seven members were white and British. In a commission which was to look into the constitutional position of India, there wasn't included a single Indian. As it happened, Lord Sinha, an Indian life peer in the British parliament, was available and would have provided a sop to nationalist sentiments.

Birkenhead's arrogance in ignoring these feelings resulted in the commission being seen, not as a body which would give serious consideration to constitutional matters, but as an inquisition by empire builders into the fitness of Indians for self-rule.

Such a notion galvanized the Congress. It decided to boycott the commission 'at every stage and in every form'. Nationalist feelings were aroused so strongly that parties of every political affiliation came together on an anti-Simon programme.

Birkenhead's response was to set a challenge. 'I have twice in three years during which I have been secretary of state,' he said, 'invited our critics in India to put forward their own suggestions for a constitution to indicate to us the form in which in their judgment any reforms of constitution may take place. That offer is still open.'

These were the beginnings of a process which culminated in the Indian Congress' Lahore Declaration of December 1929. A series of All-Parties Conferences took place in India to draft a Constitutional Scheme. The *Nehru Report*, a project headed by Motilal Nehru, was presented at the last of the All-Parties Conferences. It advocated a parliamentary system of government under a dominion status for the country. The younger elements of the Congress wanted nothing less than complete independence. Thus there were now two camps in the Congress: the older, more careful leaders led by Motilal Nehru advocating dominion status, opposed by the younger, more radical group led by his own son Jawaharlal Nehru.

These camps were heading for a collision in the Calcutta Congress session of December 1928. Gandhi became the mediator and effected a compromise: the *Nehru Report* was adopted. The government was given a year to accept its provisions in toto. If, however, it had not been accepted by 31 December 1929 Congress would raise the stakes and demand Complete Independence. And fight for it, if necessary, by using non-violent non-cooperation.

In that one year, a great number of things happened. Industrial labour was getting militant with strikes in Bombay and Jamshedpur. Violent acts, termed 'terrorist outrages' took place in many parts of the country, the most famous of them being Bhagat Singh and B.K. Datt's bombs hurled into the Central Legislative Assembly. Not to kill anyone, they said, but 'to make the deaf hear'. A bomb had also gone off under the Viceregal train as the Viceroy neared Delhi from a trip to South India. Such signs of unrest were to be seen everywhere and violent acts were applauded throughout India, with the young anarchists being regarded as heroes and martyrs.

After that, all roads had to lead to Lahore. Especially when it was clear that whatever the nationalist sentiments, whatever the acts of violence and whatever the British concern about the law and order situation, the government wasn't going to relent on political issues, whether based on dominion status or otherwise. A lot of that had to do with the fact that even if the government in Britain was getting more liberal, there was a very strong feeling in Britain, and especially in the British parliament, against making any 'concessions' to the Indian political leadership. The Viceroy, sitting in Delhi and feeling the heat of direct action, proposed. British MPs, sitting in London and feeling the fervour of possession, disposed.

The Lahore Congress was held in the last three days of 1929. Under the presidentship of Jawaharlal Nehru (his name was put up, incidentally, by Gandhi), the Congress adopted a resolution asking for Swaraj, Complete Independence. The plan of action was to include boycotting elections and resignations from government political positions. Details were to be worked out by the All India Congress Committee (AICC) which was authorized, 'whenever it deems fit, to launch upon a programme of Civil Disobedience including non-payments of taxes'. 26 January was declared 'Purna Swaraj' (Complete Independence) Day. Purna Swaraj Day wasn't meant to be the start of Civil Disobedience. The day was marked instead by the unfurling of Independence flags and hundreds and hundreds of people throughout the country declaring:

> We believe that it is the inalienable right of the Indian people, as of any other people, to have freedom and to enjoy the fruits of their toil and have the necessities of life so that they may have full opportunities of growth. We believe also that if any government deprives a people of these rights and oppresses them, the people have a further right to alter it or to abolish it.
>
> We hold it to be a crime against man and God to submit any longer to a rule that has caused this fourfold disaster to our country. We recognize, however, that the most effective way of gaining our freedom is not through violence. We will therefore prepare ourselves,

by withdrawing, so far as we can, all voluntary association with the British Government, and will prepare for Civil Disobedience, including non-payment of taxes. We are convinced that if we can but withdraw our voluntary help and stop payment of taxes without doing violence, even under provocation, the end of this inhuman rule is assured. We therefore hereby solemnly resolve to carry out the Congress instructions issued from time to time for the purpose of establishing Purna Swaraj.

Although the All India Congress Committee was authorized to formulate a plan and put it into action, everyone knew that the onus was squarely on Gandhi's shoulders. (In fact, this was formalized at the meeting of the Congress Working Committee at Sabarmati from 14 to 16 February when Gandhi was appointed – in the strange phrase used without self-consciousness then – 'Dictator' of the future campaign.)

Did Gandhi have a plan? Would a new campaign be called off just as it was gathering steam as Gandhi did in 1922 following the killing of twenty-two policemen in Chauri Chaura? The second question bothered Congress radicals, particularly Jawaharlal Nehru. Gandhi's response came in a press interview: 'I am trying to conceive a plan whereby no suspension need take place by reason of any outside disturbance – a plan whereby Civil Disobedience once started may go on without interruption until the goal is reached.' In his weekly journal, *Young India*, he elaborated on the idea a bit further. 'To English friends,' Gandhi began his article 'that Civil Disobedience may resolve itself into violent disobedience is, I am sorry to have to confess, not an unlikely event. But I know that it will not be the cause of it. Violence is there already corroding the whole body politic. Civil Disobedience will be but a purifying process and may bring to the surface what is burrowing under and into the whole body.'

As to the first question of whether he had formulated a plan, his reply to Rabindranath Tagore was frank. 'I am furiously thinking night and day. But I do not see any light coming out of the surrounding darkness.'

For Gandhi, the problem stemmed from a reality everyone shied away from but which he, in his usual frank way, enunciated: 'It is a gross misrepresentation of the true situation to say that the masses are impatient to be led to Civil Disobedience, and that I am hanging back. I know well enough how to lead to Civil Disobedience a people who are prepared to embark upon it on my terms. I see no such sign on the horizon. But I live in faith.'

In spite of this gloom, Gandhi knew that he had a core group ready and prepared for action. 'If the country is able to do nothing and I see the fitness of the ashram inmates, something can certainly be done through them.'

The 'inmates' of Sabarmati Ashram were certainly ready for the Civil Disobedience Movement. The regime they were subjected to was tough and unrelenting. They were the men and women closest to Gandhi, certainly in the physical sense, and being close to Gandhi was certainly not the most comfortable place to be in.

THE ASHRAMITES OF SABARMATI

When Gandhi returned from South Africa with a group of eighteen, which included both relatives and followers, he wanted to set up a base similar to his Phoenix settlement.

He received invitations to set up an ashram from many parts of the country as varied as Rajkot, Calcutta and Hardwar. Gandhi chose Ahmedabad, probably for pragmatic reasons. The first reason was that it was a major city with a lot of the advantages that go with it. A second possible reason was that some local industrialists had offered to provide funds to set up and run the ashram. A third reason could have been that the city was a textile centre and hand-spinning and weaving were close to Gandhi's heart. And lastly, Gujarat, of course, was Gandhi's home state and if he was to come back from a foreign country, surely it should be to a place where he felt really at home.

Called the Satyagraha Ashram, it started off with a population of twenty-five men, women and children staying in a bungalow donated by a local lawyer. It was, at best, a temporary solution to the needs of the satyagrahis and soon Gandhi was looking for a more appropriate site. He found it in a plot of wooded land covering thirty-six acres on the bank of the River Sabarmati. Later Sabarmati Ashram grew into a 150-

acre settlement containing cottages, a school, a common dining room, a library and sheds for spinning and weaving. There was space also for a dairy farm and for growing vegetables and cotton. 'An ashram without orchard, farm or cattle would not be a complete unit,' Gandhi wrote.

Gandhi liked the Sabarmati site for other reasons: it was near the Dadheechi temple (and Dadheechi was revered for his renunciation of worldly goods); it was near the Dudheshwar cremation grounds (a constant reminder of human mortality); it was near the Sabarmati Central Jail (jail-going was supposed to be the occupational hazard of a satyagrahi).

Everything seemed to be settling down on this idyllic site, but soon there was trouble in paradise. A family of 'untouchables' wanted to join the ashram: Dudabhai, a teacher in Bombay, his wife Danibehn and baby daughter Lakshmi were more than willing to adhere to the ashram's strict discipline. Gandhi of course, said yes. All hell broke loose.

So many years later, we can scarcely imagine the prejudices that were deeply ingrained in the India of that time. The wealthy businessmen and industrialists who supported the ashram financially were outraged and cut off their funding. Gandhi's answer was to announce that he would move into an Untouchable colony in the direst slums of Ahmedabad and earn a living through manual work. Before he could do that, the ashram received an anonymous donation of the then considerable sum of Rs 13,000 (the unnamed donor turned out to be Ambalal Sarabhai of the well-known Sarabhai family), and Gandhi did not have to carry out his threat.

There was, however, revolt within the ashram too. And, much to Gandhi's discomfiture, one of the 'rebels' was Kasturba, his own wife. Gandhi's way of dealing with this nascent rebellion was characteristic: he decided to make an example of the person closest to him. Kasturba either accepted everyone in the ashram as equal, including an Untouchable. Or she quit. And came back when she had got over her prejudices. She stayed.

The life Dudabhai and Danibehn embraced wasn't an easy one as this daily routine will show.

4 a.m.	Wake-up bell
4.15 to 4.45 a.m.	Morning prayer
5 to 6.10 a.m.	Bath, exercise, study
6.10 to 6.30 a.m.	Breakfast
6.30 to 7 a.m.	Women's prayer class
7 to 10.30 a.m.	Manual labour, education and sanitation work
10.45 to 11.15 a.m.	Lunch
11.15 to 12 noon	Rest
12 to 4.30 p.m.	Manual labour and classes
4.30 to 5.30 p.m.	Recreation
5.30 to 6 p.m.	Dinner
6 to 7 p.m.	Recreation
7 to 7.30 p.m.	Common worship
7.30 to 9 p.m.	Recreation
9 p.m.	Retiring bell

This routine was strictly enforced. Lapses were noted; three lapses and you were out of the ashram. The most common failings were being late for prayer meetings and not spinning the daily allotted quota of yarn. At prayer meetings, the gates were shut exactly on time. Then the rolls were called and each ashramite had to announce how much yarn had been spun. Someone pointed out that this wasted time: why not earmark each of the three gates for a specified quota of yarn and let people record their figures in ledgers as they walked in? This saved a half hour. Gandhi was pleased.

There was a spiritual underpinning to the physical discipline, which came from Gandhi's reading while in prison in South Africa. He was profoundly influenced by Henry David Thoreau (1817-62), the American writer who became well-known for his attacks on many contemporary social institutions and for his faith in the religious significance of nature. In his essay 'Civil Disobedience' (1849), Thoreau said that people should refuse to obey any law they believed was unjust. 'There will never be a really free and enlightened State,' he wrote, 'until the State comes to recognize the individual as a higher and independent power, from which all its own power and authority are derived, and treats him accordingly.' Thoreau's *Walden* (1854) records his observations of nature and his life when he moved to the shore of Walden Pond near Concord, Massachusetts. The book, a celebration of people living in harmony with nature, was the inspiration of Gandhi's life in an ashram.

The other two writers who influenced Gandhi were John Ruskin, the nineteenth century English writer whose four essays in *Unto This Last* (1862), were critiques of the free enterprise system and mass-produced goods, and Leo Tolstoy, the great Russian writer and social reformer, particularly Tolstoy's essay 'The Kingdom of God is Within You' (1894). In the essay Tolstoy said that people can affirm the goodness in themselves if they carry out a rigorous self-examination and are willing to reform themselves. Tolstoy also believed that any use of violence is to be deplored and that the use of force should be opposed non-violently.

From these influences, refined through his own sensibilities and experiences, emerged Gandhi's philosophy of Satyagraha. The aim of the ashram was to live self-sufficiently with nature and by following certain precepts, become a true satyagrahi.

These precepts bear a startling resemblance to those enunciated by the ancient sage Patanjali, though Patanjali's name does not get mentioned in any Gandhi bibliography. Perhaps hundreds of years on, they had become part of a universal consciousness. These precepts were *satya* (truth), *ahimsa* (non-violence), *aparigraha* (non-materialism), *astey* (non-covetousness) and *brahmacharya* (celibacy). These weren't just theoretical concepts; they were practical aids to daily living.

Surprisingly for a man with a reputation for rigidity, Gandhi's interpretation of these precepts was pragmatic rather than doctrinaire. Thus *satya* had no fixed formula. 'There is nothing wrong,' he said, 'in everyone following truth according to his lights.'

The principle of ahimsa has many interpretations. Rajendra Prasad in his autobiography gives one closest to Gandhi's: 'If one is afraid of one's adversary, and desists from harming him, that is not non-violence. One who desists out of fear would unhesitatingly attack his adversary if he gets the latter at a disadvantage... Only he can be called non-violent, who be he strong or weak, desists from causing hurt to others because of his belief that it is wrong to hurt others... such behaviour is possible only when convinced of the justice of one's cause.'

The concept of *brahmacharya* may seem puritanical as many of Gandhi's ideas on sexuality appear to be, but at the centre of it was the idea of restraint. Of discipline. These applied not just to sexual matters but also to diet, to social service, to manual labour and even to prayer.

Adherence to these disciplines, Gandhi felt, were essential prerequisites to achieve the moral and emotional controls required of a true satyagrahi.

As Gandhi prepared for the 'event' (whose final shape had still not formed in his mind), he wrote down for his would-be satyagrahis a comprehensive 'rule book' of Satyagraha.

SOME RULES OF SATYAGRAHA

> Satyagraha literally means insistence of truth. This insistence arms the votary with matchless power. This power or force is connoted by the word Satyagraha. Satyagraha, to be genuine, may be offered against parents, against one's wife or one's children, against rulers, against fellow citizens, even against the whole world.
>
> Such a universal force necessarily makes no distinction between kinsmen and strangers, young and old, man and

woman, friend and foe. The force to be so applied can never be physical. There is in it no room for violence. The only force of universal application can, therefore, be that of ahimsa or love. In other words it is soul force.

Love does not burn others, it burns itself. Therefore, a satyagrahi, i.e., a civil resister will joyfully suffer even unto death.

It follows, therefore, that a civil resister, whilst he will strain every nerve to compass the end of the existing rule, will do no intentional injury in thought, word or deed to the person of a single Englishman. This necessarily brief explanation of Satyagraha will perhaps enable the reader to understand and appreciate the following rules.

AS AN INDIVIDUAL

1. A satyagrahi, i.e., a civil resister will harbour no anger.
2. He will suffer the anger of the opponent.
3. In so doing he will put up with assaults from the opponent, never retaliate; but he will not submit, out of fear of punishment or the like, to any order given in anger.
4. When any person in authority seeks to arrest a civil resister, he will voluntarily submit to the arrest, and he will not resist the attachment or removal of his own property, if any, when it is sought to be confiscated by authorities.
5. If a civil resister has any property in his possession as a trustee, he will refuse to surrender it, even though in defending it he might lose his life. He will, however, never retaliate.
6. Non-retaliation excludes swearing and cursing.
7. Therefore a civil resister will never insult his opponent, and therefore also not take part in many of the newly coined cries which are contrary to the spirit of ahimsa.
8. A civil resister will not salute the Union Jack, nor will he insult it or officials, English or Indian.
9. In the course of the struggle if any one insults an official or

commits an assault upon him, a civil resister will protect such official or officials from the insult or attack even at the risk of his life.

AS A PRISONER

10. As a prisoner, a civil resister will behave courteously toward prison officials, and will observe all such discipline of the prison as is not contrary to self-respect; as for instance, whilst he will Salaam officials in the usual manner, he will not perform any humiliating gyrations and refuse to shout 'Victory to Sarkar' or the like. He will take cleanly cooked and cleanly served food, which is not contrary to his religion, and will refuse to take food insultingly served or served in unclean vessels.
11. A civil resister will make no distinction between an ordinary prisoner or himself, will in no way regard himself as superior to the rest, nor will he ask for any conveniences that may not be necessary for keeping his body in good health and conditions. He is entitled to ask for such conveniences as may be required for his physical or spiritual well-being.
12. A civil resister may not fast for want of conveniences whose deprivation does not involve any injury to one's self-respect.

AS A UNIT

13. A civil resister will joyfully obey all the orders issued by the leader of the corps, whether they please him or not.
14. He will carry out orders in the first instance even though they appear to him insulting, inimical or foolish, and then appeal to higher authority. He is free before joining to determine the fitness of the corps to satisfy him, but after he has joined it, it becomes a duty to submit to its discipline

irksome or otherwise. If the sum total of the energy of the corps appears to a member to be improper or immoral, he has a right to sever his connection, but being within it, he has no right to commit a breach of its discipline.
15. No civil resister is to expect maintenance for his dependents. It would be an accident if any such provision is made. A civil resister entrusts his dependents to the care of God. Even in ordinary warfare wherein hundreds of thousands give themselves up to it, they are able to make no previous provision. How much more then, should such be the case in Satyagraha? It is the universal experience that in such times hardly anybody is left to starve.

IN COMMUNAL FIGHTS

16. No civil resister will intentionally become a cause of communal riots.
17. In the event of any such outbreak, he will not take sides, but he will assist only that party which is demonstrably in the right. Being a Hindu he will be generous towards Musalmans and others, and will sacrifice himself in the attempt to save non-Hindus from a Hindu attack. And if the attack is from the other side, he will not participate in any retaliation but will give his life in protecting Hindus.
18. He will, to the best of his ability, avoid every occasion that may give rise to communal quarrels.
19. If there is a procession of satyagrahis they will do nothing that would wound the religious susceptibilities of any communities, and they will not take part in any other processions that are likely to wound such susceptibilities.

In the weekly journal *Young India* which he edited and which he used as a forum for his thoughts and, often, to give instructions, he wrote on 27 February 1930 about what his followers should do when he was arrested:

Let us, however, think of the immediate future. This time on my arrest there is to be no mute, passive non-violence, but non-violence of the activest type should be set in motion, so that not a single believer in non-violence as an article of faith for the purpose of achieving India's goal should find himself free or alive at the end of the effort to submit longer to the existing slavery. It would be, therefore, the duty of every one to take up such Civil Disobedience or civil resistance as may be advised and conducted by my successor, or as might be taken up by the Congress. I must confess, that at the present moment, I have no all-India successor in view. But I have sufficient faith in the co-workers and in the mission itself to know that circumstances will give the successor. This peremptory condition must be patent to all that he must be an out-and-out believer in the efficacy of non-violence for the purpose intended. For without that living faith in it he will not be able at the crucial moment to discover a non-violent method.

It must be parenthetically understood that what is being said here in no way fetters the discretion and full authority of the Congress. The Congress will adopt only such things said here that may commend themselves to Congressmen in general. If the nature of these instructions is to be properly understood, the organic value of the charter of full liberty given to me by the Working Committee should be adequately appreciated. Non-violence, if it does not submit to any restrictions upon its liberty, subjects no one and no institution to any restriction whatsoever, save what may be self-imposed or voluntarily adopted. So long as the vast body of Congressmen continue to believe in non-violence as the only policy in the existing circumstances and have confidence not only in the bona fides of my successor and those who claim to believe in non-violence as an article of faith to the extent indicated but also in the ability of the successor wisely to guide the Movement, the Congress will give him and them its blessings and even give effect to these instructions and his.

So far as I am concerned, my intention is to start the Movement only through the inmates of the ashram and those who have submitted to its discipline and assimilated the spirit of its methods. Those, therefore, who will offer battle at the very commencement will be unknown to fame. Hitherto the ashram has been deliberately kept in reserve in order that by a fairly long course of discipline it might acquire stability. I feel, that if the Satyagraha Ashram is to deserve the great confidence that has been reposed in it and the affection lavished upon it by friends, the time has arrived for it to demonstrate the qualities implied in the word Satyagraha. I feel that our self-imposed restraints have become subtle indulgences, and the prestige acquired has provided us with privileges and conveniences of which we may be utterly unworthy. These have been thankfully accepted in the hope that some day we would be able to give a good account of ourselves in terms of Satyagraha. And if at the end of nearly fifteen years of its existence, the ashram cannot give such a demonstration, it and I should disappear, and it would be well for the nation, the ashram and me.

When the beginning is well and truly made I expect the response from all over the country. It will be the duty then of every one who wants to make the Movement a success to keep it non-violent and under discipline. Every one will be expected to stand at his post except when called by his chief. If there is a spontaneous mass response, as I hope there will be, and if previous experience is any guide, it will largely be self-regulated. But every one who accepts non-violence whether as an article of faith or policy would assist the mass movement. Mass movements have, all over the world, thrown up unexpected leaders. This should be no exception to the rule. Whilst, therefore, every effort imaginable and possible should be made to restrain the forces of violence, Civil Disobedience once begun this time cannot be stopped and must not be stopped so long as there is a single civil resister left free or alive. A votary of Satyagraha should find himself in one of the following states:

1. In prison or in an analogous state; or
2. Engaged in Civil Disobedience; or
3. Under orders at the spinning wheel, or at some constructive work advancing Swaraj.

In a sense, the inmates of Sabarmati Ashram were being trained by Gandhi much as a commander would train an army. But while the latter would impart knowledge of warfare and enforce the required discipline to wage it, Gandhi's little army was being trained in a far more difficult form of warfare. Here truth was to prevail. The enemy was not to be regarded as an enemy but as a fellow human being who had to be won over. The enemy's very real weapons of violence were to be disregarded and faced without fear even though one had no weapons of one's own except a belief in one's cause ... It was tough being a satyagrahi. It was tough being an ashramite. Which is why when Gandhi planned his major offensive of Civil Disobedience, he knew that his soldiers in the battle would all be from Sabarmati Ashram.

THE CHOICE OF WEAPONS

Why salt? Why Dandi? Why a march? Complex questions to which there are no simple answers.

There were many reasons given for Gandhi's decision to use salt as his main weapon. Some of them were quite fanciful. Victor Wolfenstein wrote:

> Another line of interpretation, which is consonant with the view I have been developing of Gandhi's personality, is suggested by Ernest Jones' contention that one of the two basic symbolic significances of salt is human semen. If it had this unconscious meaning for Gandhi, then we may understand his depriving himself of condiments, including salt, as a form of sexual abstinence, involving a regression to an issue of the oral phase. In the context of Salt March, Gandhi's taking of salt from the British can thus be seen as reclaiming for the Indian people the manhood and potency which was properly theirs. The British had denied the Indians their manhood by monopolizing the manufacture of salt; Gandhi, in order to break that monopoly and restore the virility of the Indian people, risked punishment and death – an altruistic act, for as just noted, Gandhi himself had renounced the use of salt (that is, was celibate).

As it happens, Gandhi had written extensively about salt, his earliest essay on the subject being one of his first articles. He wrote equally extensively about sexual matters. The connections he made had to do with salt and food as well as salt and health. There was no connection made, in any of his articles, between salt and sex. Of course, he could have made a subconscious connection. Who can tell? But even Eric Erikson, who wrote a controversial psycho-analytical study of Gandhi and analyzed his sexual writings in the minutest detail, didn't go along with the Wolfenstein idea. 'Anybody acquainted with the ancient Indian preoccupation with semen as a substance which pervades the whole body and which, therefore, is released only at the expense of vitality, activity and spiritual power, will have to admit that if there is an equation between salt and semen in the primitive mind, the Indian people more than any other could be assumed to make the most of it.' He went on to say, tongue in cheek perhaps, that 'in any context except that of irrationality clearly attributable to sexual repression, one should take any interpretation that explains a human act by recourse to sexual symbolism with a grain of salt.'

But why go into the subconscious? With Gandhi, every move he made and every decision that he took was thought out, articulated, written up and discussed. He had the intuition of a genius but not the arrogance. He was particularly brilliant at finding the simplest, pared down symbol which would instantaneously strike a chord with just about everyone. There was the charkha, the common, unfashionable, nearly discarded spinning wheel made into a powerful symbol of mass empowerment. There was the loin cloth, an ungainly, inconvenient, bare cover for nakedness, resonating with the undress of the underprivileged and in vivid contrast to the Plus Fours of the Empire's rulers. There was the insistence on third class rail travel, another powerful symbol identifying him with a large swathe of the population which knew of no other form of travel (and which too was in ironical contrast with the pomp and pageantry of the Viceroy's horse-drawn carriages and special saloon-car trains). And then there was Sabarmati Ashram…

This would suggest that Gandhi's decision wasn't based on some convoluted subconscious connection. It was based on the simple, direct connection between salt and the poor. 'There are millions in India who live upon one pice – one-third of a penny – a day,' wrote a young Gandhi in the London journal *The Vegetarian Messenger* of 1 June 1891, well before he went to South Africa and well before he became Mahatma Gandhi. 'These poor people have only one meal per day, and that consists of stale bread and salt, a heavily taxed article.' Thus the pre-Mahatma, youthful lawyer in England was already concerned about the injustice of a tax on salt. At Sabarmati nearly forty years later, the injustice must have loomed rather large.

What was particularly galling for any nationalist was the fact that India did not import any salt before the British set up shop and decided they wanted only the British variety. Liverpool salt was dumped into India, a Salt Commission was set up in 1885 which recommended in an oblique kind of way that local salt should be taxed to make the import of British salt more attractive. The Salt Act which followed gave the government the monopoly on the manufacture of salt, imposed heavy taxes on local manufacture and well-nigh killed the local industry.

The iniquity of the tax regime was referred to again and again by Gandhi in many of his writings:

> Next to air and water, salt is perhaps the greatest necessity of life. It is the only condiment of the poor. Cattle cannot live without salt. Salt is a necessary article in many manufactures. It is also a rich manure.
>
> There is no article like salt outside water by taxing which the State can reach even the starving millions, the sick, the maimed and the utterly helpless. The tax constitutes therefore the most

inhuman poll tax that the ingenuity of man can devise. The wholesale price per maund of 82 lbs. is according to government publications as low as 10 pies, and the tax, say, 20 annas, i.e., 240 pies. This means 2,400 per cent on sale price! What this means to the poor can hardly be imagined by us. Salt production like cotton growing has been centralized for the sake of sustaining the inhuman monopoly. The necessary consequence of the wilful destruction of the spinning wheel was destruction of cottage cultivation of cotton. The necessary consequence of salt monopoly was the destruction, i.e., closing down of salt works in thousands of places where the poor people manufactured their own salt. A correspondent writes to me from Konkan, saying that if the people had freedom, they could pick up salt from the deposits made by the receding tides on the bountiful coast. But he sorrowfully adds the officers turn the salt over into the sea as fast as nature deposits it. He adds however, that those who can successfully evade the salt police do help themselves to this sea salt. Gujarat workers report the existence of many places where, but for the prohibition, people can get their salt as easily as they can dig out earth for many household purposes. Free Bengal can today manufacture all the salt she can ever need. And yet she is forced to import all the salt she eats.

Here is what a retired salt officer writes without disclosing his name:

"Under the law the manufacture of salt includes every process by which salt is separated from brine or earth or any other liquid or solid substances and also every process for the purification of refinement of salt.

"Contraband salt means salt or salt earth which has not paid duty.

1. Manufacture, removal, or transport of salt without licence;
2. The excavation, collection, or removal of natural salt or salt-earth;

3. And possession or sale of contraband salt are punishable with a fine up to Rs 500 or imprisonment up to six months or both.

"The whole western littoral of the Bombay Presidency from Cambay to Ratnagiri; the whole cast of Kathiawad and the southern coast of Sindh is a huge natural salt work, and natural salt and salt earth from which salt can be easily prepared is in every creek.

"If a band of volunteers begin the work all along the coast, it would be impossible for the whole strength of the police and customs staff to prevent them from collecting natural salt and salt earth, turning them into salt in the interior and retailing it. The people of the presidency or at least the men and women of the older generation firmly believe, that locally manufactured salt is healthier than Kharaghoda salt, and they would love to have it, while every one would like to have cheap salt. The poor people on the coast will join in the collection of salt spontaneously in these days of unemployment. Trying to get salt from government salt works without paying duty would be stealing or robbery, an act of First Class Hinsa that would justify even shooting down of the offenders if they persisted in the act."

I have given the letter as it was received. When salt can be manufactured much more easily than it can be taken from salt depots, I am not likely to advise people to help themselves to the article from salt pans or storehouses. But I do not share the salt officer's characterization of such helping as first class *himsa*. Both the helping from pans and manufacturing contraband salt are statutory crimes heavily punishable. Why is the manufacturing without licence a virtue and taking salt from a manufacturing pan a vice? If the import is wrong, it is wrong whether in connection with manufactured salt or the crude article. If a robber steals my grain and cooks some of it, I am entitled to both the raw and the cooked grain. I may draw a distinction for the sake of avoiding inconvenience between manufactured and crude salt, and adopt the easier method of manufacturing salt. But that does not alter the legal position in the slightest degree. When therefore the time comes, civil resisters will have an ample opportunity of their ability to conduct their campaign regarding the tax in a most effective manner. The illegality is in a government that steals the people's salt and makes them pay heavily for the stolen article. The people, when they become conscious of their power, will have every right to take possession of what belongs to them.

In retrospect, the choice of breaking the salt laws seems obvious, especially when you consider that the alternatives were to focus on Chaukidari tax, the Land Revenue tax, the Forest Laws and the Grazing Areas Laws. Yet at that time, there was no real consensus that breaking of the salt laws would catch the public imagination. To many, it didn't quite fit the bill for a lofty cause like Total Independence. People like Motilal Nehru felt that this was yet another example of the 'food faddist' Gandhi. Indulal Yajnik, a Gujarati politician with radical leanings, asked, 'Wouldn't the Salt Campaign fail to arouse the enthusiasm of the youth of the nation? Wouldn't they all see through the farce of wielding the sledgehammer of Satyagraha to kill the fly of the Salt Act?'

Other ideas emerged: the burning of foreign cloth (an old Gandhi idea) on a mass scale was one. Boycott of law courts was another. Non-payment of land tax was appealing but Gandhi feared that the government would use that as an excuse to seize property and not return it. Jawaharlal Nehru suggested forming a parallel government, which was rejected because it might have become no more than a debating club. The man of action, Vallabhbhai Patel wanted a mass march to Delhi.

But Gandhi's mind was made up. He wanted a symbol which was universal, something the poorest peasant would understand and identify with. He also saw the easy accessibility of salt throughout the country as a big plus as it would enable a large number of people to break the law.

A lot of what men do has its origins in their own past. In his articles in *Young India*, as well as in his conversations with other leaders before the march, Gandhi maintained that he was sure that the salt tax would be the focus of the campaign, but he wasn't sure what form it would take. 'I had not the ghost of a suspicion how the breach of Salt Law would work itself out,' he was to say later. 'Pandit Motilalji and other friends were fretting and did not know what I would do; and I could tell them nothing, as I myself knew nothing about it. But like a flash it came.'

The flash said the campaign should be in the form of a march. And surely the inspiration came from Gandhi's own past; in fact, his not-too-distant past in South Africa.

There on 6 November 1913 Gandhi led a group of Indians across the Transvaal border in a march to protest against a whole clutch of highly discriminatory laws, which included compulsory registration (likened then to a 'dog collar') for all Indians who had come to South Africa as indentured labourers. Another law invalidated all marriages not solemnized in a church. Yet another forbade Indians from crossing the Transvaal border.

That's just what Gandhi did, in what became his first organized march and his first notable success with Satyagraha. Gandhi had earlier thought that his 'Army of Peace' would consist of 'at least sixteen, at most sixty-six satyagrahis'. In the event, 2037 men, 127 women and fifty-seven children took part in the 'illegal' crossing. The *Sunday Post* reported, 'The pilgrims whom Gandhi is guiding are an exceedingly picturesque crew. To the eye they appear most meagre, indeed emaciated; their legs are mere sticks but the way they are marching on the starvation rations show them to be particularly hardy.' One particular incident will show how hardy these poor labourers were, and just how determined. One of the 127 women was carrying a small baby. While crossing a stream, she stumbled and fell and the baby drowned. 'I don't want to grieve over the dead. I want to work for those who are alive,' she said and marched on.

Walking was also a habit which Gandhi had encouraged years earlier. In his ashram called Tolstoy Farm, one of the rules made it compulsory for the inmates to use no transport when they went to the nearest city, Johannesburg. So what if Johannesburg was over thirty kilometres away? For the inmates it was a necessary weekly journey. The 'walker' woke up at two in the morning, and half an hour later, in the pitch dark, began his journey, which would take six to seven hours. (The record, incidentally, was four hours eighteen minutes!) Work done, the walker would return in the evening.

There is yet another facet to walking in a group which must have struck Gandhi fairly early on. Note the words of the *Sunday Post* report: 'The pilgrims whom Gandhi is guiding...' the story began. That is something Gandhi would have seized on, the idea of a pilgrimage, so close to the religious Indian mind. Kedarnath, Badrinath, Haji Malang, Vaishnodevi, Ambaji... These shrines, built at different periods in history, built by different sets of people, built by different sects in different parts of the country, had one thing in common: they weren't for the faint-hearted. The journeys are always tough, tests of physical endurance and willpower, in their hardship an examination of the worshipper's resolve. The way to your goal is never easy; the harder it is, the worthier, and more virtuous the journey. In the hardship there is atonement, there is propitiation and ultimately, redemption. Gandhi's marchers in Transvaal were true pilgrims in that sense, and the future marchers of Sabarmati Ashram too would soon follow that path.

There they were then, all the elements now coming together: salt, the poor man's essential condiment; the breaking of an unjust law; the courage of Satyagraha to take on violence without flinching; the pilgrims' walk for salvation...

THE GIRDING OF LOINS

When Satyagraha is your weapon of choice for the coming battle, you better forget every known rule of warfare. For example, discard the notion of taking your enemy by surprise. Which is why Gandhi wrote to the Viceroy on 2 March 1930, a full ten days before he was to set off on his march, and some days before he had worked out the details in his own mind.

Discarding the usual obsequious forms of address of those days, especially to the highest authority in the land, Gandhi chose the simple 'Dear Friend'. As the text shows, friendly the letter certainly wasn't particularly where it deals with British economic policies in India. Gandhi, half-apologetically, even gets personal, dragging into the discussion the question of the Viceroy's rather handsome salary. Gandhi himself wrote to his ashramites, 'The letter... is not an ultimatum, but it is a friendly, if also a frank, communication from one who considers himself to be a friend of Englishmen.'

DEAR FRIEND
Before embarking on Civil Disobedience and taking the risk I have dreaded to take all these years, I would fain approach you and find a way out.

My personal faith is absolutely clear. I cannot intentionally hurt anything that lives, much less fellow human beings, even though they may do the greatest wrong to me and mine. Whilst, therefore, I hold the British rule to be a curse, I do not intend harm to a single Englishman or to any legitimate interest he may have in India.

I must not be misunderstood. Though I hold the British rule in India to be a curse, I do not, therefore, consider Englishmen in general to be worse than any other people on earth. I have the privilege of claiming many Englishmen as dearest friends. Indeed much that I have learnt of the evil of British rule is due to the writings of frank and courageous Englishmen who have not hesitated to tell the unpalatable truth about the rule.

And why do I regard the British rule as a curse?

It has impoverished the dumb millions by a system of progressive exploitation and by a ruinously expensive military and civil administration which the country can never afford.

It has reduced us politically to serfdom. It has sapped the foundations of our culture. And, by the policy of cruel disarmament, it has degraded us spiritually. Lacking the inward strength, we have been reduced, by all but universal disarmament, to a state bordering on cowardly helplessness.

In common with many of my countrymen, I had hugged the fond hope that the proposed Round Table Conference might furnish a solution. But, when you said plainly that you could not give any assurance that you or the British Cabinet would pledge yourselves to support a scheme of full Dominion Status, the Round Table Conference could not possibly furnish the solution for which vocal India is consciously, and the dumb millions are unconsciously, thirsting. Needless to say there never was any question of parliament's verdict being anticipated. Instances are not wanting of the British Cabinet, in anticipation of the parliamentary verdict, having pledged itself to a particular policy.

The Delhi interview having miscarried, there was no option for Pandit Motilal Nehru and me but to take steps to carry out the solemn resolution of the Congress arrived at in Calcutta at its session in 1928.

THE TREND OF BRITISH POLICY

But the Resolution of Independence should cause no alarm, if the word Dominion Status mentioned in your announcement had been used in its accepted sense. For, has it not been admitted by responsible British statesmen, the Dominion Status is virtual Independence? What, however, I fear is that there never has been any intention of granting such Dominion Status to India in the immediate future.

But this is all past history. Since the announcement many events have happened which show unmistakably the trend of British policy.

It seems as clear as daylight that responsible British statesmen do not contemplate any alteration in British policy that might adversely affect Britain's commerce with India or require an impartial and close scrutiny of Britain's transaction with India. If nothing is done to end the process of exploitation India must be bled with an ever increasing speed. The Finance Member regards as a settled fact the 1/6 ratio which by a stroke of the pen drains India of a few crores. And when a serious attempt is being made through a civil form of direct action, to unsettle this fact, among many others, even you cannot help appealing to the wealthy landed classes to help you to crush that attempt in the name of an order that grinds India to atoms.

Unless those who work in the name of the nation understand and keep before all concerned, the motive that lies behind the craving for Independence, there is every danger of Independence itself coming to us so charged as to be of no value to those toiling voiceless millions for whom it is sought and for whom it is worth taking. It is for that reason that I have been recently telling the public what Independence should really mean.

WHAT INDEPENDENCE MEANS

Let me put before you some of the salient points.

The terrific pressure of land revenue, which furnishes a large part of the total, must undergo considerable modification in an Independent India. Even the much vaunted permanent settlement benefits the few rich Zamindars, not the ryots. The ryot has remained as helpless as ever. He is a mere tenant at will. Not only, then, has the land revenue to be considerably reduced, but the whole revenue system has to be so revised as to make the ryot's good its primary concern. But the British system seems to be designed to crush the very life out of him. Even the salt he must use to live is so taxed as to make the burden fall heaviest on him, if only because of the heartless impartiality of its incidence. The tax shows itself still more burdensome on the poor man when it is remembered that salt is the one thing he must eat more than the rich man both individually and collectively. The drink and drug revenue, too, is derived from the poor. It saps the foundation both of their health and morals. It is defended under the false plea of individual freedom, but, in reality, is maintained for its own sake. The ingenuity of the authors of the reforms of 1919 transferred this revenue to the so-called responsible part of dyarchy, so as to throw the burden of prohibition on it, thus, from the very beginning, rendering it powerless for good. If the unhappy minister wipes out this revenue he must starve education, since in the existing circumstances he has no new source of replacing that revenue. If the weight of taxation has crushed the poor from above, the destruction of the central supplementary industry, i.e., hand-spinning, has undermined their capacity for producing wealth. The tale of India's ruination is not complete with reference to the liabilities incurred in her name. Sufficient has been recently said about these in the public press. It must be the duty of a free India to subject all the liabilities to the strictest investigation, and repudiate those that may be adjudged by an impartial tribunal to be unjust and unfair.

STAGGERING PHENOMENON

The iniquities sampled above are maintained in order to carry on a foreign administration, demonstrably the most expensive in the world. Take your own salary. It is over Rs 21,000 per month, besides many other indirect additions. The British Prime Minister gets £ 5000 per year, i.e., over Rs 5400 per month at the present rate of exchange. You are getting over Rs 700 per day against India's average income of less than annas 2 per day. The Prime Minister gets Rs 180 per day against Great Britain's average income of nearly Rs 2 per day. Thus you are getting much over five thousand times India's average income. The British Minister is getting only ninety times Britain's average income. On bended knee I ask you to ponder over this phenomenon. I have taken a personal illustration to drive home a painful truth. I have too great a regard for you as a man to wish to hurt your feelings. I know that you do not need the salary you get. Probably the whole of your salary goes for charity. But a system that provides for such an arrangement deserves to be summarily scrapped. What is true of the Viceregal salary is true generally of the whole administration.

THAT EMBRACE OF DEATH

A radical cutting down of the revenue, therefore, depends upon an equally radical reduction in expenses of the administration. This means a transformation of the scheme of government. This transformation is impossible without Independence. Hence, in my opinion, the spontaneous demonstration of 26 January, in which hundreds of thousands of villagers instinctively participated. To them Independence means deliverance from the killing weight.

Not one of the great British political parties, it seems to me, is prepared to give up the Indian spoils to which Great Britain helps herself from day to day, often, in spite of the unanimous opposition of Indian opinion.

Nevertheless, if India is to live as a nation, if the slow death by starvation of her people is to stop, some remedy must be found for immediate relief. The proposed conference is certainly not the remedy. It is not a matter of carrying conviction by argument. The matter resolves itself into one of matching forces. Conviction or no conviction, Great Britain would defend her Indian commerce and interest by all the forces at her command. India must consequently evolve force enough to free herself from that embrace of death.

SINFUL TO WAIT ANY LONGER

It is common cause that, however disorganized, and, for the time being, insignificant, it may be, the party of violence is gaining ground and making itself felt. Its end is the same as mine. But I am convinced that it cannot bring the desired relief to the dumb millions. And the conviction is growing deeper and deeper in me that nothing but unadulterated non-violence can check the organized violence of the British Government. Many think that non-violence is not an active force. My experience, limited though it undoubtedly is, shows that non-violence can be an intensely active force. It is my purpose to set in motion that force as well against the organized violent force of the British rule as the unorganized violent force of the growing party of violence. To sit still would be to give rein to both the forces above mentioned. Having an unquestioning and immovable faith in the efficacy of non-violence, as I know it, it would be sinful on my part to wait any longer.

This non-violence will be expressed through Civil Disobedience, for the moment confined to the inmates of the Satyagraha Ashram, but ultimately designed to cover all those who choose to join the Movement with its obvious limitations.

MY AMBITION – CONVERSION OF BRITISH PEOPLE

I know that in embarking on non-violence I shall be running what might fairly be termed a mad risk. But the victories of truth have never been won without risks, often of the gravest character. Conversion of a nation that has consciously or unconsciously preyed upon another, far more numerous, far more ancient and no less cultured than itself, is worth any amount of risk.

I have deliberately used the word conversion. For my ambition is no less than to convert the British people through non-violence, and thus make them see the wrong they have done to India. I do not seek to harm your people. I want to serve them even as I want to serve my own. I believe that I have always served them up to 1919 blindly. But when my eyes were opened and I conceived non-cooperation, the object still was to serve them. I employed the same weapon that I have in all humility successfully used against the dearest members of my family. If I have equal love for your people with mine it will not long remain hidden. It will be acknowledged by them even as the members of my family acknowledged it after they had tried me for several years. If the people join me as I expect they will, the sufferings they will undergo, unless the British nation sooner retraces its steps, will be enough to melt the stoniest heart.

IF YOU CANNOT SEE YOUR WAY

The plan through Civil Disobedience will be to combat such evils as I have sampled out. If we want to sever the British connection it is because of such evils. When they are removed the path becomes easy. Then the way to friendly negotiation will be open. If the British commerce with India is purified of greed, you will have no difficulty in recognizing your independence. I respectfully invite you then to pave the way for immediate removal of those evils, and thus open a way for a real conference between equals, interested only in promoting the common good of mankind through voluntary fellowship and in arranging terms of mutual help and commerce equally suited to both. You have unnecessarily laid stress upon the communal problems that unhappily affect this land. Important though they undoubtedly are for the consideration of any scheme of Government, they have little bearing on the greater problems which are above communities and which affect them all equally. But if you cannot see your way to deal with these evils and my letter makes no appeal to your heart, on the eleventh day of this month, I shall proceed with such co-workers of the ashram as I can take, to disregard the provisions of the salt laws. I regard this tax to be the most iniquitous of all from the poor man's standpoint. As the Independence Movement is essentially for the poorest in the land the beginning will be made with this evil. The wonder is that we have submitted to the cruel monopoly for so long. It is, I know, open to you to frustrate my design by arresting me. I hope that there will be tens of thousands ready, in a disciplined manner, to take up the work after me, and, in the act of disobeying the Salt Act to lay themselves open to the penalties of a law that should never have disfigured the Statute book.

NO THREAT BUT A SACRED DUTY

I have no desire to cause you unnecessary embarrassment, or any at all, so far as I can help. If you think that there is any substance in my letter, and if you will care to discuss matters with me, and if to that end you would like me to postpone publication of this letter, I shall gladly refrain on receipt of a telegram to that effect soon after this reaches you. You will, however, do me the favour not to deflect me from my course unless you can see your way to conform to the substance of this letter.

This letter is not in any way intended as a threat but is a simple and sacred duty peremptory on a civil resister. Therefore I am having it specially delivered by a young English friend who believes in the Indian cause and is a full believer in non-violence and whom Providence seems to have sent to me, as it were, for the very purpose.

I remain,
Your sincere friend
M.K.Gandhi

Lord Irwin didn't deign to reply. Instead Gandhi got this curt response:

Dear Mr Gandhi,

His Excellency the Viceroy desires me to acknowledge your letter of the 2nd March. He regrets to learn that you contemplate a course of action which is clearly bound to involve violation of the law and danger to the public peace.

Yours very truly,
G. Cunningham
Private Secretary

Gandhi reacted in some anguish but with no surprise. 'On bended knee,' he said, 'I asked for bread and I have received stone instead.'

He saw the rebuff as final justification for battle. He wrote:

It was open to the Viceroy to disarm me by freeing the poor man's salt, tax on which costs him 5 annas per year or nearly three day's income. I do not know outside India any one who pays to the State Rs 3 per year if he earns Rs 360 during that period. It was open to the Viceroy to do many other things except sending the usual reply. But the time is not yet. He represents a nation that does not easily give in, that does not easily repent.

Entreaty never convinces it. It readily listens to physical force. It can witness with bated breath a boxing match for hours without fatigue. It can go mad over a football match in which there may be broken bones. It goes into ecstacies over bloodcurdling accounts of war. It will listen also to mute resistless suffering. It will not part with the millions it annually drains from India in reply to any argument, however convincing. The Viceregal reply does not surprise me.

But I know that the salt tax has to go and many other things with it, if my letter means what it says. Time alone can show how much of it was meant.

The reply says I contemplate a course of action which is clearly bound to involve violation of the law and danger to the public peace. In spite of the forest of books containing rules and regulations, the only law that the nation knows is the will of the British administrators, the only public peace the nation knows is the peace of a public prison. India is one vast prison house. I repudiate this law and regard it as my sacred duty to break the mournful monotony of the compulsory peace that is choking the heart of the nation for want of free vent.

There is a little parenthesis to the sending of Gandhi's letter to the Viceroy to which he alludes in the last paragraph. This little story illustrates, once again, that every action of Gandhi, however small, was premeditated and well thought out. Gandhi didn't send the Viceroy's letter by post. He chose a special messenger.

The messenger, he decided, had to be English. Madeline Slade, better known as Miraben had become too identified with India's cause. Gandhi chose Reginald Reynolds, a young Englishman who had come to India some months earlier and was staying at the ashram. Gandhi explained his choice:

For me sending of the letter was a religious act as the whole struggle is. And I selected an English friend as my messenger, because I wanted to forge a further check upon myself against any intentional act that would hurt a single Englishman. If I have a sense of honour in me, this choice should prove an automatic restraint even upon unconscious error. It pleases me also to have the unselfish and unsolicited association of a cultured, well read, devout Englishman in an act which may, in spite of all my effort to the contrary, involve loss of English life.

Thus Gandhi had used Reynolds, not just to send his letter to the Viceroy, but also to send a message to the world that it wasn't just Indians who were joined in a peaceful battle against the British; even the British identified with the Indian cause.

That wasn't all. Gandhi didn't want Reynolds to be just a courier. To truly identify with the cause, he needed to know the contents and agree with them. 'Read it,' Gandhi said and waited for Reynold's response. The young man, no doubt, was overwhelmed at the importance he was being given.

The letter was delivered on 4 March. In spite of the widespread speculation about its contents (and a fair amount of second-guessing and rumours), it wasn't made public until 6 March so that Lord Irwin had time to reply.

The plan of action was revealed in a prayer meeting at the ashram. Gandhi would lead a march starting on 12 March with a group of fifty in 'the direction of Pethapur', final destination unknown. There the group would pick up salt and carry it away and thus break the law.

The response was electric. As Jawaharlal Nehru said in a telegram to Gandhi, 'The news was fully expected and yet it sent a thrill through all of us. How I wish I would join your gallant band!' It seemed everyone from all parts of the country wanted to join in. Letters and telegrams poured in. Political leaders, ex-ashramites, ordinary men and women, students willing to give up their exams, they all wanted to be part of the first group of marchers.

Gandhi, however, was clear. He didn't want a huge march for one simple reason: a large group would be difficult to control. His one single worry was that, given extreme provocation, someone from the group would break rank and get violent. That little spark could set off a conflagration.

That is why Gandhi was certain that every marcher would be hand-picked by him. He was also certain that each of the hand-picked marchers would be schooled in the concept of Satyagraha to such an extent that it would be an intrinsic part of him. The only people who fitted this particular bill were members of the Sabarmati Ashram, men not only dedicated to the cause, but who had volunteered to adhere to the rigorous discipline of the ashram. They would, without doubt, accept the march's hardships. They would also, like true satyagrahis, not rise to any provocative bait dangled before them by enforcers of the law.

For Gandhi, the rigours and hardships of the march were a necessary feature. In 1929 he had propagated the use of khadi by walking tours in Andhra and UP (then United Provinces). He was horrified to be greeted by welcoming arches. He was mortified that he was put up in 'luxuriously fitted rooms'. Worse, everything in those rooms was foreign, including the linen!

He now wanted no scope for misunderstanding. He published his 'request' so that there would be no foreign cloth, no arches, no nothing:

> I requested the mahajans and the workers of the respective places to bear in mind the following. The satyagrahi party is expected to reach each place by 8 o'clock in the morning and to sit down for lunch between 10.00 and 10.30 a.m. It may be half past nine by the time the party reaches Aslali on the first day. No rooms will be needed for rest at noon or night, but a clean, shaded place will be enough. In the absence of such shaded place, it will be enough to have a bamboo-and-grass covering. Both bamboo and grass can be put to use again.
>
> It is assumed that the village people will provide us food.
> If provisions are supplied, the party will cook its own meal.

The food supplied, whether cooked or uncooked, should be the simplest possible. Nothing more than rotli or rotla or kedgeree with vegetables and milk or curds, will be required. Sweets, even if prepared, will be declined. Vegetables should be merely boiled, and no oil, spices and chillies, whether green or dry, whole or crushed should be added or used in the cooking. This is my advice for preparing a meal:

Morning before departure	*Rab* and *dhebra*; the *rab* should be left to the party itself to prepare.
Midday	*Bhakhri,* vegetable and milk or buttermilk.
Evening, before the march is resumed	Roasted gram, rice.
Night	Kedgeree with vegetable and buttermilk or milk.

The ghee for all the meals together should not exceed three *tolas* per head: One *tola* in the *rab*, one served separately to be smeared on the *bhakhri*, and one to be put into the kedgeree. For me goat's milk, if available, in the morning, at noon and at night, and raisins or dates and three lemons will do. I hope that the village people will incur no expenses whatever, except for the simple food items named above.

Gandhi sent out a further set of instructions for villages to adhere to as he passed through them on his march:

I look forward to meeting the people of each village and its neighbourhood.
　　Everyone in the party will be carrying his own bedding, so

that the village people have to provide nothing except a clean place for resting in.

The people should incur no expense on account of betel leaves, betel-nuts or tea for the party.

I shall be happy if every village maintains complete cleanliness and fixes beforehand an enclosed place for the satyagrahis to answer calls of nature. If the villagers do not already use khadi, it is clear that they should now start using it.

It is desirable that information under the following heads should be kept for each village:

1. Population: Number of women, men, Hindus, Muslims, Christians, Parsis, etc.
2. Number of Untouchables.
3. If there is a school in the village, the number of boys and girls attending it.
4. Number of spinning wheels.
5. The monthly sale of khadi.
6. Number of people wearing khadi exclusively.
7. Salt consumed per head; salt used for cattle, etc.
8. Number of cows and buffaloes in the village
9. The amount of land revenue paid; at what rate per acre?
10. The area of common grazing-ground if any.
11. Do the people drink? How far is the liquor shop from the village?
12. Educational and other facilities, if any, for the Untouchables.

It will be good if this information is written out on a sheet of paper neatly and handed to me immediately on our arrival.

This formidable litany of self-denial didn't deter the ashramites, all of whom wrote out a letter of application stating why they wanted to be part of the march, and be in the very first batch. Many of these applicants were women.

He surprised them with a 'No'. This march would have only men. But weren't women the equal of men? Hadn't Gandhiji himself insisted that everyone in the ashram was equal? Why did he, now, discriminate? Did he think that women couldn't cope with hardships?

Gandhi's reply once again showed that everything he did had a reason. In this case, he said, 'It would be cowardice to take women with us,' because their presence amongst the marchers would inhibit British law-enforcers. The British normally did not attack women, so taking women with him would, in a sense, be like hiding behind them, using them as a shield. Gandhi had never needed a shield. A true satyagrahi should also not need one.

Not everyone – especially those who were outsiders – was as enthusiastic about the idea of this long march. We already know of Motilal Nehru's reservations. The pro-British press, ridiculed the idea of the march, finding it almost laughable. *The Statesman's* comments were typical of the British press:

> Mr Gandhi has revealed his secret. His scheme of Civil Disobedience is to go with some of his followers to the seashore to take water from the sea and extract salt from it by evaporation. It is difficult not to laugh, and we imagine that will be the mood of most thinking Indians. There is something almost childishly theatrical in challenging in this way the salt monopoly of the government. It is to be hoped that none of the peasants will be induced to use salt obtained in this crude manner, for the effects could prove rather alarming. Let Mr Gandhi and his followers eat their own salt, and they are likely to be disqualified from political agitation for some time without any intervention on the part of the government.

In retrospect, it seems unbelievable that so many people failed to see the viscerally emotional appeal of the Salt March. If only they could have been at Sabarmati Ashram...

THE MEDIUM AND THE MESSAGE

Mahatma Gandhi used two main means of communication at the ashram. One was the speech he would give at the daily prayer meeting. He would, depending on the circumstances, reprimand or cajole, inform or instruct. Very often, he would use this pulpit to inspire his flock.

The second form of communication was to write articles. Gandhi saw himself as a bit of a journalist, and there is a book there for the writing on the many uses he made of his writing skills. Quite often it was to reiterate to ashramites something he had already spoken about. Occasionally, it was to issue a clarification or an elaboration. He also used his articles to reach a wider audience far beyond the walls of the ashram: he used them to connect with his followers throughout the country. Or to answer an opponent (or to shame him by agreeing with his wildest arguments). He also used his articles to reach British authorities and inform them of his intentions and convey a rationale for his future actions.

The publication he used most often at this time was *Young India*, published from Ahmedabad, whose masthead proclaimed 'A Weekly Journal, Edited by M.K. Gandhi'. A single copy cost 2 annas, a six-month subscription Rs 3 and an annual subscription Rs 5. The masthead also made it clear that this no frills, no pictures, no sops newspaper knew of its appeal to a wider readership; foreign subscription was listed at Rs 7, 12 shillings or $3.

As with everything, Gandhi had definite views about journalism. In an issue of *Young India* soon after his letter to the Viceroy had been sent, Gandhi pulled up the press:

> My letter to the Viceroy went on the 2nd instant as anticipated by the newspapers. Forecasts have been published of its contents which are largely untrue. I wish these correspondents and the news agencies will, instead of making the publication of news a matter merely of "making money", think of the public good. If there had been anything to give to the public, surely Pandit Jawaharlal Nehru would have given it. But it was thought advisable to wait for an acknowledgement from Delhi before publishing the letter. I am not intent on a fight. I am leaving no stone unturned to avoid it. But I am ready for it the moment I find that there is no honourable way out of it. Premature publication of news indirectly obtained by means not always straight ought not to be the function of journalists. I know that the newspaper said to be the greatest in the world makes it a boast to obtain by secret methods news which no other agency can. It makes it a boast to publish news which the keepers are most anxious often in the public interest to withhold for the time being. But the English public submits to the treatment, because monied and influential men conduct the *Times*. We have blindly copied the rulers' code of manners without discrimination in the matter of publication of news as in many others of still greater importance. I know that mine is a voice in the wilderness, though I speak with the authority of an unbroken experience of practical journalism for over twenty years, if successful conducting of four weeklies can be regarded as such. Be that as it may, the imminent fight includes among the points of attack this slavish habit of copying everything

English. No one will accuse me of any anti-English tendency. Indeed I pride myself on my discrimination. I have thankfully copied many things from them. Punctuality, reticence, public hygiene, independent thinking and exercise of judgment and several other things I owe to my association with them. But never having had the slightest touch of slave mentality in me and never having even a thought of materially benefiting myself through contact, official or otherwise with them, I have had the rare good fortune of studying them with complete detachment. On the eve of battle therefore I would warn fellow journalists against copying the English method of obtaining and publishing news. Let them study my original method which was introduced long before I became a Mahatma and before I had acquired any status of importance in the public life of India. It was a hard struggle, but I found in the field of journalism as in many others that the strictest honesty and fair dealing was undoubtedly the best policy. Any shorter cut is longer at least by double the length sought to be saved. For there must be a retracing. I say all this not for the sake of reading a lesson to fellow journalists but for the sake of the struggle in which I would value the cooperation of journalists whether they approve of or oppose my methods of political warfare. Let them not add to the risks I am already taking. The rule I would like them to observe is never to publish any news without having them checked by some one connected with me and having authority.

Through the pages of *Young India*, Gandhi carried out a kind of guerilla warfare with the pro-British Indian press, mainly *The Times of India* and *The Statesman*. There was also the pro-British *British* press to contend with. As preparations for the Dandi March were in progress, a special issue of the London *Times* was brought out in England. It needed a rebuttal. But for once, it wasn't Gandhi writing a reply; it was young Reginald Reynolds. No doubt Gandhi felt it would be more appropriate if an Englishman were to take issue with an English publication.

Reynold's pen dripped with the heavy sarcasm of a young man new to journalism:

> A truly amazing publication lies before me as I write. It is the London *Times* of 18 February 1930. But it is more than that : it is the Special Indian number, on which the *Times*' staff has been working for over a year; so you may be sure that it is solid good stuff.
>
> Seldom has Fleet Street produced such a galaxy of learning all in one issue. Not only has His Excellency the Viceroy given it his blessing, but the number contains articles from the pens of such well-known Indians as Sir Harcourt Butler, the Marquis of Zetland, Sir Valentine Chirol, Sir William Marris, Bishop Palmer, Sir Walter Lawrence and Lord Inchcape, as well as quite a host of army officers.
>
> The *Times* has indeed produced a complete encyclopaedia of Indian life. As Lord Irwin says, it has "given the British people a picture of India". Here you may learn all about the beneficent rule of the East India Company and the wise Government of India today. Here you will find the reason why the Simon Commission had to be composed of "God's Englishmen". Here you may read about the splendid police service of India, how unimpeachable is the system of justice, and how corrupt and rotten every "Native" institution is. There are pictures, too, of Big Game Hunting, the Viceroy's House, Simla, a Maharaja's state elephant, the New India House (London), the Calcutta Cathedral, Sir William Birdwood, the Viceroy and another Lord (with wives), the Simon Commission, a P and O Liner and other things that concern the Indian peasant.
>
> But what interested me most was that part which dealt with Indian trade and finance.
>
> "Excess of Exports over Imports", "Thriving Home Industries" – Those were the headlines that especially caught my eye.
>
> According to this writer the fact is that ever since the British

came to India, India has apparently been growing richer and richer at the expense of England. "We know of no time," says the *Times* correspondent, "when the balance of trade was not in India's favour against Europe."

Now that is a very amazing statement which we shall do well to examine. In the first place we know that the East India Company had a monopoly of Anglo-Indian trade till the year 1793. Now the trading system of the East India Company was naively piratical. They taxed their subjects, and, after deducting the expenses of administration, remitted the balance to England partly in cash and partly in goods which were bought with Indian money in the Indian market for sale in other parts of the world. This was known as the Company's investment.

Henry Verelst, Governor of Bengal, estimated that in the three years 1766-68 the Company's imports totalled £624,375 whilst its exports reached the figure of £6,311,250. But he was by no means deceived by this "favourable trade balance". In a letter, dated September 1767, quoted by Romesh Dutt in his *Economic History*, Verelst wrote: "Each of the European Companies, by means of money taken up in the country, have greatly enlarged their annual investments, without adding a rupee to the riches of the provinces."

Till 1833 the Company continued this policy of buying goods in India at the expense of the Indian tax payer and selling them outside the country for the profit of the English shareholder. In that year their charter was renewed on condition of their giving up trade and contenting themselves with the spoils of administration. The "favourable trade balance", however, still continued, and that for an obvious reason. Not only had the Company's "profits" still to be met out of the revenues, not only were salaries and other expenses largely disbursed in England, but the Company had further borrowed money to the tune of over thirty million sterling, chiefly for war purposes, i.e., totally unproductive expenditure. The story of those iniquitous debts and how India became saddled with them has been recently told in several articles in the Indian press. I am concerned, however, rather with their effects than with their origin.

From now on the Indian Trade Balance had to cope with an ever increasing drain as both debts and cost of administration rose steadily. To point now to the excess of exports which is required to meet this annual tribute and speak of it as evidence of prosperity is either an extremely indecent form of humour or something that is called by an uglier name. In point of fact, most of this excess of export represents so much clear loss to the Indian peasant. Under foreign rule India has sunk from one of the richest countries in the world to the poorest; and out of her poverty she still pays a tribute of precious grain, that rich men from England may draw fat salaries from her exchequer.

I have been turning over the pages of the *Times* again. On every page, one finds the same kind of stuff, written by baronets and business peers with an air of pompous finality. The lies themselves we are used to; what worries me is that thousands of people in England will have read them and believed them.

Under such circumstances, what hope is there of justice descending like manna from England? I speak with experience when I say that the Englishman is fed on lies so far as India is concerned right from the time when he first opens a school history textbook. The Tories are in a solid phalanx against India. Mr Lloyd George is raving almost daily in the press about the impossibility of Indian self-government. The Labour Party talks in bland promises and have so far behaved like the Tories.

One way only lies open to India, and Gandhiji has already taken the first step. He knows better than any one the futility of hoping that England will own herself to be in the wrong. I admit with shame that moral cant and hyprocrisy have become a sort of political second nature with us. The crowning farce of India's ignominy would be for her representatives to "give evidence" regarding their own affairs before our self-appointed tribunal. And Lord Irwin has repeatedly made it clear that that is the meaning of the so-called "Round Table Conference".

England is reading about an India that is prosperous and happy beneath the aegis of British rule, where "agitation" is confined to a handful of lawyers and students. The Lloyds and O'Dwyers and Butlers are telling them every day how greatly the Indian peasant rejoices in the blessings of the British Raj, not till the nation rises in its masses against them will these countrymen of mine understand. For their sakes and ours I pray to God that Revolution may come at the call of Gandhiji; for come it must by one of two ways, and the other will deluge two nations in blood.

Gandhi often used a device familiar to journalists to clarify his intentions and actions to an audience both inside and outside the ashram. This device was the Q&A. A format, where Gandhi posed the questions which he imagined were uppermost in the minds of both follower and detractor. Some days before announcing his Dandi plans, he used this format again to reach 'friends as well as critics':

Q. Surely you are not so impatient as to start your campaign without letting the authorities know your plans and giving them an opportunity for meeting you and arresting you?

A. Those who know my past should know that I hold it to be contrary to Satyagraha to do anything secretly or impatiently. My plans will be certainly sent to the Viceroy before I take any definite step. A satyagrahi has no secrets to keep from his opponent or so-called enemy.

Q. Did you not say even at Lahore that the country was not prepared for Civil Disobedience, especially, no-tax campaign on a mass scale?

A. I am not even now sure that it is. But it has become clear to me as never before that the unpreparedness in the sense that a non-violent atmosphere is wanting will as time goes by, very likely increase as it has been increasing all these years. Young men are impatient. I know definitely many stayed their violent designs because in 1921 the Congress had decided to offer Civil Disobedience. That school has been more active than before because of my repeated declarations that the country was not prepared for Civil Disobedience. I feel then that if non-violence is an active force, as I know it is, it should work even in the face of the most violent atmosphere. One difficulty in the way was that the Congress claiming to represent the whole nation could not very well offer Civil Disobedience and disown responsibility for violence especially by Congressmen. I have procured discharge from that limitation by taking over the responsibility of launching on Civil Disobedience. I represent no one but myself and at the most those whom I may enrol for the campaign. But I propose at present to confine myself only to those who are amenable to the ashram discipline and have actually undergone it for some time. It is true that I may not shirk responsibility indirectly for any violence that may break out on the part of the nation and in the course of the campaign. But such responsibility will always be there and can be only a degree more than the responsibility I share with the British ruler in their sins against the nation in so far as I give my cooperation however reluctantly and ever so slightly. For instance I give my cooperation by paying taxes direct or indirect. The very salt I eat compels my voluntary cooperation. Moreover it has dawned on me never so plainly as now that if my non-violence has suffered the greatest incarnation of violence which the British imperialistic rule is, it must suffer the crude and ineffective violence of the impatient patriots who know not that by their ineffectiveness they are but helping that imperialistic rule and enabling it to consolidate the very thing they seek to destroy. I see now as clearly as daylight that my non-violence work as it has done against the British misrule has shaken it somewhat. Even so will it shake the counter-violence of the patriot if taking courage in both my hands I set my non-violence actively in motion, i.e., Civil

Disobedience. I reduce the risk of the outbreak of counterviolence to a minimum: by taking sole charge of the campaign. After all is said and done, however, I feel the truth of the description given to my proposal by *The Times of India*. It is indeed "the last throw of a gambler". I have been a "gambler" all my life. In my passion for finding the truth and in relentlessly following out my faith in non-violence, I have counted no stake too great. In doing so I have erred, if at all, in the company of the most distinguished scientist of any age and any clime.

Q. But what about your much vaunted faith in Hindu-Muslim unity? Of what value will even Independence be without that unity?

A. My faith in that unity is as bright as ever. I do not want Independence at the cost even of the weakest minority, let alone the powerful Musalman and the no less powerful Sikh. The Lahore Congress resolution on unity finally sums up all its previous effort on that behalf. The Congress rules out all solutions proposed on a communal basis. But if it is ever compelled to consider such a solution it will consider only that, which will give (not merely justice) but satisfaction to all the parties concerned. To be true to its word therefore, the Congress cannot accept any scheme of Independence that does not give satisfaction, so far as communal rights are concerned, to the parties concerned. The campaign that is about to be launched is calculated to generate power for the whole nation to be independent. But it will not be in fact till all the parties have combined. To postpone Civil Disobedience which has nothing to do with communalism till the latter is set at rest will be to move in a vicious circle and defeat the very end that all must have in view. What I am hoping is that the Congress being free from the communal incubus will tend it, if it remains true to the nation as a whole, to become the strongest centre party jealously guarding the rights of the weakest members. Such a

Congress will have only servants of the nation, not office-seekers. Till Independence is achieved or till unity is reached it will have nothing to do with any office or favours from the government of the day in competition with the minorities. Happily the Congress has now nothing to do with the legislatures which have perhaps more than anything else increased communal bitterness. It is no doubt unfortunate that at the present moment the Congress contains largely only the Hindu element. But if the Congress Hindus cease to think communally and will take no advantage that cannot be shared to the full with all the other communities, it will presently disarm all suspicion and will attract to itself the noblest among Musalmans, Sikhs, Parsis, Christians, Jews and all those who are of India. But whether the Congress ever approaches this ideal or not, my course is, as it always has been, perfectly clear. This unity among all is no new love with me. I have treasured it, acted up to it from my youth upward. When I went to London as a mere lad in 1889 I believed in it as passionately as I do now. When I went to South Africa in 1893 I worked it out in every detail of my life. Love so deep seated as it is in me will not be sacrificed even for the realm of the whole world. Indeed this campaign should take the attention of the nation off the communal problem and to rivet it on the things that are common to all Indians no matter to what religion or sect they may belong.

Q. Then you will raise, if you can, a force ultimately hostile to the British?

A. Never. My love for non-violence is superior to every other thing mundane or supramundane. It is equalled only by my love for truth. My scheme of life if it draws no distinction between different religionists in India, also draws none between different races. I embark upon the campaign as much out of my love for the Englishman as for the Indians. By self-suffering I seek to convert him, never to destroy him.

Q. But may not all this be your hallucination that can never come to pass in this matter of fact world of ours?

A. It may well be that. It is not a charge wholly unfamiliar to me. My hallucinations in the past have served me well. This last is not expected to fail me. If it does, it will but harm me and those who may come or put themselves under its influence. If my hallucination is potent to the authorities, my body is always at their disposal. If owing to my threatened action any Englishman's life is put in greater danger than it is now, the arm of English authority is long enough and strong enough to overtake any outbreak that may occur between Kashmir and Cape Comorin or Karachi and Dibrugarh. Lastly no campaign need take place, if all the politicians and editors instead of addressing themselves to me will address themselves to the authorities and ask them to undo the continuing wrongs some of which I have inadequately described in these pages.

The propaganda war continued this way right through the Salt Tax campaign. It was an unequal battle: the big newspapers were all British-owned and thus pro-British; the nationalist newspapers reached only a committed readership. Newspapers in England were, of course, solidly for Empire.

As we will see, one of the effects of the Dandi Salt March was to begin a process of change in the attitude of the press. This was to play a far-reaching role in influencing world opinon.

THE FINAL COUNTDOWN

Before the final countdown to the Dandi Salt March began, events occurred in Sabarmati Ashram which, in any other place, would have caused an upheaval. But such was the discipline at the ashram, so unwavering was the loyalty to Gandhi and so complete was the acceptance of Gandhian principles, that life went on in the normal, everyday kind of way.

What did happen was that smallpox, then very much an active and contagious viral disease, hit the ashram. Three children died. That in itself wasn't unusual for the 1920s and 1930s. What followed was: Mourning and any demonstration of grief, an inmate of Sabarmati Ashram notes, 'were out of the question'. But that wasn't enough. 'All, including the parents,' the account continues, 'were to go through the days' duties as though nothing has happened. And all stood the test well.'

Apparently, only the 'minimum necessary' number of people went for the cremation and there was no break in the work allotted to the rest. Amazingly, even the parents of the children, it was said, 'did not miss their prayers or their spinning'.

This had a lot to do not just with the work ethic but also an avowed refusal to make any distinction between what would normally be regarded as an occasion for grief or an occasion for rejoicing. 'Within two days of the death of a boy, came the day (previously fixed) of a wedding of a girl. It was gone through with all the solemnity that a sacred rite requires, and Gandhiji spoke on the restraining quality of marriage with as much fervour as on the benignant quality of death.'

This is remarkable enough. But there's more. And that springs from the other Gandhi, the one who had views and beliefs that would be regarded today as eccentric, even obscurantist. Gandhi did not believe in vaccination. In fact, all his life he had been a staunch opponent of the procedure most people regarded as routine and life-saving. This did not mean that ashramites could not get themselves or their families vaccinated, just that Gandhi did nothing to persuade them to do so. That also meant that he himself refused to be vaccinated, in spite of entreaties from everywhere that he should do so, given the grave risk of infection.

At the evening prayer meeting, he said:

> How can I go back on the principles I have held dear all my life, when I find that it is these principles that are being put to the test? I have no doubt in my mind that vaccination is a filthy process, that it is harmful in the end and that it is little short of taking beef. I may be entirely mistaken. But holding the views that I do, how can I recant them? Because I see child after child passing away? No, not even if the whole of the ashram were to be swept away, may I insist on vaccination and pocket my principle. What would my love of truth and my adherence to principle mean, if they were to vanish at the slightest touch of reality?
>
> ... But God is putting me through a greater test. On the eve of what is to be the final test of our strength, God is warning me through the messenger of death. I have tried hydropathy and earth treatment with success in numerous cases. Never has the treatment failed as it seems to have done during the month. But does that mean that I must therefore lose faith in the treatment

and faith in God? Even so my faith in the efficacy of non-violence may be put to the severest test. I may have to see not three but hundreds and thousands being done to death during the campaign I am about to launch. Shall my heart quail before that catastrophe, or will I persevere in my faith? No, I want you everyone to understand that this epidemic is not a scourge, but a trial and preparation, a tribulation sent to steel our hearts and to chain us more strongly and firmly to faith in God. And would not my faith in the Gita be a mockery if three deaths were to unhinge me? It is as clear to me as daylight that life and death are but phases of the same thing, the reverse and obverse of the same coin. In fact tribulation and death seem to me to present a phase far richer than happiness or life. What is life worth without trials and tribulations which are the salt of life. The history of mankind would have been a blank sheet without these individuals. What is *Ramayana* but a record of the trials, privations and penances of Rama and Sita. The life of Rama, after the recovery of Sita, full of happiness as it was, does not occupy even a hundredth part of the epic. I want you all to treasure death and suffering more than life, and to appreciate their cleansing and purifying character.

Notwithstanding that sobering advice, it was life, in the sense of the future course of action, rather than death, as in the past, which was occupying Gandhi's attention. As well as, of course, other leaders like Sardar Vallabhbhai Patel and Jawaharlal Nehru.

It was clear to them that once it was decided that Civil Disobedience would take the shape of a Salt March led by Gandhi, there were three elements which would be key to their success. The first was timing. The second the choice of marchers and the third the route of the march.

The timing was more or less decided: it had to be now, and the culmination of the march had to coincide with the beginning of National Week commencing on 6 April. The identity of the people to take part in the march was no longer an issue: Gandhi had decided that only his ashramites had the discipline and stamina required. It was only a question of whittling down the numbers from the two hundred occupants of the ashram to the fifty or so that Gandhi thought would be a manageable crowd.

The last, and vital, element of the march was the route and its final destination. The route had to be such that it created the maximum impact; the final destination had to be such as to be picturesque to provide the ideal photo op (even though that phrase had not then been invented). Most important of all, the route had to traverse politically sympathetic areas.

At this distance, and given the mythology that has grown up around it, it is easy to forget that in the early 1930s, the freedom struggle was not one cohesive, integrated and popular movement, involving most of the country. There was, as there is now, and as always will be, a general political apathy. British rule may have been economically exploitative, but it wasn't cruel and oppressive. Many saw it as comfortable and benign; many others saw it as more measured and democratic (and certainly far less whimsical) than the rule of maharajas and nawabs. As with most revolutionary movements, the leadership was provided by the intelligentsia and the upper middle class, and they, in fact, formed the main body of Indian nationalists. But to succeed, the group's support base had to have at least the appearance of a mass movement, firstly to put pressure on the British and secondly to act as the catalyst which would help it to really achieve mass action. In other words, wherever Gandhi and his fellow marchers stopped, and Gandhi gave his speech, his audience needed to be enthusiastic and his audience needed to be large.

In order for that to happen, the route had to be planned in such a way that it traversed areas where Gandhi's support was strongest. Between Ahmedabad and the coast, two areas were identified: Kheda district and Bardoli, a taluka of Surat district.

Both areas were, in a manner of speaking, already 'indoctrinated' with the idea of Satyagraha. The Kheda

Satyagraha of 1918 was already the stuff of folklore: when the crops failed that year, the Patidar farming community asked the local government to suspend revenue assessment for the year. When the request was rejected, Gandhi suggested Satyagraha. The outcome wasn't completely successful in Gandhi's terms, but concessions were made by the authorities. More importantly, the poorest Patidars had seen – for the first time in their lives – that they weren't as helpless as they had thought. Satyagraha had empowered them.

Yet another plus from the Kheda Satyagraha was that Vallabhbhai Patel, a bright, young lawyer practicing in and around the area, joined Gandhi's movement. As a matter of fact, Patel went on to lead the Bardoli Satyagraha of 1928 (the issue, again, was land revenue) and was, therefore, the right man in the right place at the right time.

Patel made a significant difference to the Dandi March. First of all, he used his immense organizational skills to prepare the ground for the march. Second, it was he who suggested a much longer march than was first envisaged. The first choice of Badalpur on the coast as the final destination was ruled out at the last minute precisely because of that. Since it was only 125 kilometres from the ashram, the distance would be covered in only eight days. Patel wanted a 25-day walk to allow momentum to build up and so get maximum publicity.

The committee (including Patel) to select the site considered various options: Tithal in Surat district as well as Dharasana, Udvada (better known as a place of pilgrimage for Parsis) and Dihen. Jalalpur was considered favourably but in the end Dandi it was, on the seaside but nothing more than a village. Its claim to fame was that the Bharat Vidyalaya at Karadi, a school which was a great supporter of Gandhi, was nearby. Most importantly, it had a large flat area where the tide came in and then went out leaving behind a thin film of sea water to dry. These acted as natural salt pans and the locals for years had collected salt for their own use from there.

Even in 1930 it was becoming clear that a sizeable chunk of the Muslim population of India, though behind the nationalist cause, weren't Congress supporters. They were particularly suspicious of Gandhi because of his frequent resort to religious symbolism (which was all Hindu), his frequent quotations from the Gita and his daily prayer meetings. Gandhi's campaigns were being seen, not as nationalist campaigns but as Hindu nationalist campaigns. Muslim leadership had also rejected the Motilal Nehru Report which had said no to reserved Muslim seats in legislatures. These were some of the factors which made Muslim leaders distance themselves from Gandhi's Civil Disobedience Movement, however hard he tried to persuade them that it was completely secular. There is a telling quote from Mahomed Ali, a Muslim leader of that period: 'We refuse to join Mahatma Gandhi because his movement is not a movement for the complete independence of India but for making the seventy million of Indian Musalmans dependent on the Hindu Mahasabha.'

This had a direct impact on the choice of route to Dandi. If a straight line were to be drawn for the route from Sabarmati to Dandi, the marchers would have gone through areas where the percentage populations of Muslims would be higher than the national average. For example, some Ahmedabad districts had a thirteen per cent Muslim population and in Broach district it reached a high of twenty-four per cent.

The solution was to bypass the high-density Muslim areas and take the route through high-density Patidar localities. For example, in Broach district, the march bypassed Hansot which had a fifty-one per cent Muslim population and instead went through Mangarol and Umracchi, which were much smaller centres, but had a majority of Patidars. Yet another reason for this kind of re-routing was that Gandhi expected many local officials to resign from their government jobs as a mark of protest against the British. The Patidar officials were likely to do so and Gandhi wanted to receive their resignations in person.

Sardar Patel, who was mobilizing informal cadres along the route could not do much to change this support system even if he wanted to: his own supporters were Patidars, while amongst Muslims, his stock was not very high.

Patel had been given a dual role. He was to be the chief mobilizer and organizer of the march, the leader of the advance troops so to speak. And he was to take over as a leader of the march as soon as the inevitable arrest of Gandhi took place.

Events have a habit of not going according to plan, however. As part of the advance preparations, Patel had been going from one town to another, from village to village, giving speeches, educating people on Satyagraha and Civil Disobedience as well as exhorting them to support the forthcoming march in large numbers. In today's parlance, his message was *Jail bharo, rasta nahi roko.*

On 7 March, just five days before Gandhi was to set off from Sabarmati, Patel was arrested at a village called Ras as he began one of his speeches. The arrest was ordered by the Collector of Kheda; just the previous day he had served an order on Patel not to give a public speech for a month. Patel was sentenced to three months' rigorous imprisonment plus a penalty of Rs 500. Incidentally, the government in Bombay did not want Patel arrested and sent a cable to the Collector to that effect. But the telegram arrived after the event.

As soon as he was arrested, Patel took leave of the crowd with this short speech:

> I hope that you will hail with rejoicings the honour that has fallen to my lot of being the first to go to jail on the eve of this campaign, and that Gujarat will fulfil what the Congress and Gandhiji have expected of it.
>
> Let not Gujarat forget that the time has now arrived for repaying what Gandhiji has done for it by his sustained and strenuous penance of fifteen years on the banks of the Sabarmati.

If government have a particle of sense in them, they will not lay their hands on a saintly person like him.

It is our duty to obey his command so long as he is left free. But when he also is arrested you will do what I have told you in my Broach speech.

Our victory depends entirely on our capacity for suffering and sacrifice. The progress of our work will be commensurate with the speed with which we sever all connection with this government.

It is my prayer that God may make Gujarat capable of that achievement.

Mahadev Desai gave this account of Patel's arrest:

The story of the trial and the sentence is briefly told. In a few minutes he (Patel) was taken to Borsad, the taluka headquarters, where the district magistrate had been waiting to receive this honoured accused. Some years ago the same magistrate had, as municipal commissioner of Ahmedabad, often taken his orders from the Sardar. The magistrate and the accused shook hands. What little time elapsed was taken up by the preliminaries under the Criminal Procedure Code. Then Sardar set an example to all who were to follow him by making no statement. He only said: "I plead guilty."

As they were taking him away he entrusted whatever cash and papers he had in his pocket to Sgt. Mohanlal Pandya. The magistrate who had sentenced him to simple imprisonment for three months and to a fine of Rs 500 and three weeks imprisonment in default cast wistful eyes on the cash. "How much is it?" he inquired. "Something over Rs 25," said Sgt. Pandya. "I should like to appropriate it in part payment of his fine," said the magistrate.

"No," said the Sardar. "You may not do so. It is public money."

"How am I to know it?" asked the magistrate.

"You must take me at my word," replied the Sardar, and the magistrate gladly did so.

The car was ready to take him to the Sabarmati Prison, as ready as the magistrate who served the notice and the District Magistrate who tried the Sardar.

Sabarmati jail was about four hours away. Next door, at Sabarmati Ashram everyone had just finished their prayers when they heard that Patel's car was approaching. Everyone stood in rows by the roadside to wave to him. Unexpectedly the police stopped the car for the benefit of the waiting crowd. Patel greeted Gandhi with his characteristic open laugh, which instantly became the mood of the crowd, joyful rather than sad. Patel was garlanded in the ashram style, with a yarn garland. Kumkum was applied to his forehead. Just as the car was about to move, Patel waved and smiled. 'Follow me,' he said. 'I am keeping rooms ready for you.'

Mahadev Desai later visited Patel in jail. His account conveys vividly the atmosphere of the time and the Sardar's own exuberant personality. Desai writes:

> I had the honour in company with Professor Kriplani to have the darshan of the Sardar in the jail. His first word and his last was that he was never happier in his life than at the present moment. "But you should not have deceived Bapu like that," said Professor Kriplani. "What was I to do?" said the Sardar with a merry laugh, "they deceived me. If they had told me that they were going to send me to jail I should not have gone to Borsad."
>
> "But joking apart let us know how you are being treated."
>
> "Just like an ordinary criminal. I am perfectly happy."
>
> "Don't the new Jail Rules apply to you?"
>
> "This superintendent knows nothing about the new rules, and they refused to give me a copy of the jail manual."
>
> "But let us know something about your appointments and your associates."

"Well I am in a cell which is locked up for the night at about 5.30 p.m. on week days and 3.30 p.m. on Sundays. I am afraid I might not have slept on the first day, but then there has been no difficulty. I have been sleeping like a log. But I do wish they allowed us to sleep outside in this hot weather. I think our friends who were in here in 1922 were all allowed to sleep outside."

"And food?"

"As good or bad as one can expect to get in jail. Don't bother about food. I assure you I can live on air for three months," he said, again bursting into a loud laugh.

But we pressed him for details. *Jowari* gruel was given in the morning but he did not take it for fear of getting dysentery, and then *jowari* roti and dhal or roti and vegetable every alternate day. "The gram roti is good enough for a horse," he added. He was suffering from aching teeth and I asked him how he managed to chew the *jowari* bread. "Oh, I break it up in water and get along splendidly. I tell you, you need not worry yourself about my food."

"And do you have a bed to sleep on, or a light?"

"Neither. They have given me a blanket, and Bhagvad Gita and *Tulsi Ramayana*. If I were given a light I should be able to read at night which is impossible at present."

"You want anything else to read?"

"*Ashram Bhajanavali* (hymn book) is all that I want. The three should suffice for the brief period of three months."

"And your associates?"

"Ordinary felons. Ours is called the juvenile ward, though there are in it older people than myself. They are from all parts of the country and have come in for all sorts of crimes. Our three friends from Jabalpur, sentenced for picketing, were for a day with me, but they were removed." Dangerous because familiar company, I suppose!

I then had a talk about the circumstances of his conviction and sentence, the legal or illegal aspect of which (though not the fact of it) seems to have worried some of the members of the local bar. I told him that he might be interviewed any day by some one of the vakils from Ahmedabad who contemplated to move the high court as *amicus curiae*. "Why do you bother about it?" he said. "I am quite happy here and should regret to be released earlier. As for the conviction I am quite sure that it was wrong. The magistrate was too dense to understand the law. He did not know under what section he was to convict me. He took about an hour and a half to write a judgment of eight lines."

"But did he mention the section?"

Here the jailor read out the history ticket of Prisoner Vallabhbhai Patel which showed that he was convicted under Section 71 of the Bombay District Police Act for not complying with orders under Section 54 of the same Act.

The Sardar said: "Yes it was Section 71. But I had been asked under the order not to make a speech which was calculated to do this thing or that thing. I made no speech at all. I said I was going to disobey the order and they arrested me. When the magistrate read out to me the proceedings I said, 'Why do you worry? I plead guilty. You may convict me as you like.' When he convicted me, however, he had not the courtesy to read out the judgment but simply said he was giving me the maximum sentence under the law. He saw me handing over cash and papers to Sgt. Mohanlal Pandya, and asked me what amount it was, so that he might recover part of the fine. He had not even the decency of not casting his eyes on a few rupees when the amount of fine was Rs 500. But I told him that it was public money and that he had better be careful."

He gave me a list of things that he wanted including a soap and his shaving tackle. "No razor allowed," said the superintendent, "but we shall allow you a shave."

"I know what kind of a shave you will give me," said Vallabhbhai.

But here the jailor interrupted the superintendent with evidently better knowledge of jail rules. He said: "In this case, sir, razor might be allowed, provided he does not keep it with himself. We shall give it to him when he wants."

"Quite all right," said Vallabhbhai. "But why not give me a razor and allow me to shave the others? There will be some work to my credit." And even the little parts of that inhuman machinery called the jail department could not help creaking with laughter. But they are jealous of their inhumanity, and soon repair the mistake if ever they blunder into humanity. So the jailor added: "You may have your soap, but it should not be scented soap!" And again the jail gate rang with laughter.

As we were leaving the Sardar said: "Don't worry about me. I am happy as a bird. There is only one thing over which I am rather unhappy." And he was silent for a moment. The jail superintendent and the jailor looked curiously at each other. We also wondered what it might be. "It can't be said," said the Sardar still tantalizing me. But we insisted.

"Well," said the Sardar, "one thing and one alone worries me, and that is that all the people in charge of the jail are Indians. It is through us Indians that they work this inhuman system. I wish they were foreigners, so that I might fight them. But how could I fight our own kith and kin?"

I hope the friends saw the point of the joke. As I was leaving I was rather worried that this was to be the first and the last interview that the Sardar would have, as he could have only one interview in three months.

"Oh no," he said reassuring us, and even administering a loving rebuke, "don't worry about the interviews. I don't want any one to interview me. That will only serve to remind me that the man interviewing me is still out of jail."

Patel's arrest should have been a setback for the planned march, but Gandhi, characteristically turned it to his advantage. He immediately sensed the mood of the people in general and the Patidars in particular.

'It is a good omen for us,' Gandhi said, 'that Sardar Vallabhbhai has been arrested and sentenced. It remains to be seen what use we make of this happy beginning. The fight has now commenced and we have to carry it to its conclusion. People should celebrate the Sardar's arrest and the sentence passed on him by observing a general hartal. I request the mill-owners to close the mills, the students to absent themselves from their institutions and all shopkeepers to close their shops. There should be no need to tell Gujarat to preserve peace. Our struggle must remain non-violent from beginning to end.'

On 8 March, the evening after Patel's arrest, a crowd estimated to be 50,000 strong gathered on the banks of the Sabarmati River. This must have been the precise reason why the government did not want the arrest to take place, but a combination of an overzealous local official and slow communication had given the march just the impetus that was needed.

Gandhi's speech to the large crowd was brief and to the point. He said in Gujarati:

> I had never dreamt that Sardar Vallabhbhai would be imprisoned so soon. I think his services to Gujarat, and more particularly to this city, exceed mine a hundred times. Hence it is no wonder that he has been honoured by imprisonment before me. That certainly is his good fortune and yours also. But I find myself in a difficult situation because he has been imprisoned before me. I am eager to get arrested at any cost. I want to deprive the government of its illegitimate monopoly of salt. My aim is to get the Salt Tax abolished. That is for me one step, the first step, towards full freedom. Sardar Vallabhbhai is no longer with us in this task. The people of India are now impatient and will not rest until they have won complete freedom. My voice is bound to reach the government somehow, but Gujarat should preserve complete peace. The imprisonment of Vallabhbhai is the government's way of rewarding his services in preserving complete peace during the Satyagraha at Bardoli.
>
> We have known it for years that this is the only way in which the government can reward an independent-minded and freedom-loving person like Sardar Vallabhbhai. Let us all get so

completely absorbed in our task that we win at once what we have been yearning for all these years. To fulfil the pledge we took on the 26th we should offer Civil Disobedience. Though Vallabhbhai had broken no salt law, the government had arrested him and broken my right hand, so to say. If it has imprisoned and removed one Vallabhbhai, you, the men and women of Ahmedabad, should take his place and work as his representatives. Get ready at once, if you have love for him and have come here to sacrifice yourselves. If you are ready to follow him in his self-sacrifice, we shall show to the government and to the world how our aspirations are bound to be realized. May God grant us the strength necessary for the sacrifice we have to make.

My determination to march on Wednesday morning with the first batch comprising the Ashram inmates, stands unchanged. Let everyone present here do his duty. Vallabhbhai has said in his message that his speech at Broach clearly indicates what people should do if and when Gandhi is arrested. By going to jail himself, he has been as good as his word. Let the government reward us all in that way.

Gandhi then asked everyone in the crowd to raise their hands and take a pledge:

We the citizens of Ahmedabad, men and women, hereby resolve to follow Sardar Vallabhbhai to jail, or win Complete Independence. We shall have no peace, nor will we let the government have any, till we have won Complete Independence. We believe that India's freedom is to be won through peaceful and truthful means.

I hope that the thousands of men and women present here will raise their hands and take the pledge for which I have been training the country in general, and the citizens of Ahmedabad in particular, for the last fifteen years, and which was taken at the time of the labour strike here. Raise your hands only if you have the strength to act upon it.

The response was overwhelming. Thousands of hands were raised, accompanied by loud cries of 'Mahatma Gandhi ki jai', 'Sardar Patel ki jai'.

Although Patel was the lynchpin of the march's organization, there were other elements which played a vital part. None was more important than a group of volunteers who got to be known as Arun Tukdi (literally, Dawn Brigade).

This was a small group of students from Gujarat Vidyapith, whose vice chancellor Kakasaheb Kalelkar, was a great nationalist as was its chancellor (a certain gentleman called M.K. Gandhi). Incidentally, the institute even taught a two-week course in Civil Disobedience! And shut down normal lectures and classes during the march, so that teachers and students could participate in it.

The group's name came from the fact that these were advance troops, young men who were to steal a march on the marchers for the best possible reason: to prepare the ground. If Sardar Patel's advance troops prepared the ground by mobilization and mental fine-tuning, the Arun Tukdi volunteers prepared the ground literally. It was their job to arrange the physical facilities for the marchers at all the pre-arranged stops on the route.

These 'facilities', as we have seen from Mahatma Gandhi's extremely demanding list, were spartan to the point of being ascetic. Nevertheless, someone had to make the arrangements. These were the basic ones like ensuring that there was enough drinking water plus water of some kind for bathing. There needed to be a place to sleep in and a place to pray in. Improvised kitchens had to be put in place as did that other essential, the latrines. (These may have been nothing but cordoned off trenches, but someone had to dig them.)

There was a Day Brigade and a Night Brigade, each consisting of eight young men. The former, led by Ghulam Rasul Kureshi, organized the day halts; the latter, led by Shamal Shai, organized the night halts. The groups travelled by whatever transport they could manage to get hold of: trucks, trains and tongas. Their work load was demanding in the extreme, and no one was looking after them. They were like the backup staff of an army: the glory wasn't theirs nor were the medals. Gandhi realized this at some point and called them the real satyagrahis soon after he reached Dandi.

Kalelkar's planning was meticulous. He had organized back-up troops to his backup troops, so that if any (or all) of the Arun Tukdi were arrested, another set of student volunteers would replace them. In addition, over fifty volunteers were ready to take over even from the main marchers at short notice if they were arrested.

The Congress party also swung into action. Its aim was to prepare for the one eventuality everyone expected any time, especially now that Patel was in jail: the arrest of Mahatma Gandhi. Without losing the focus of the Salt Law in the Civil Disobedience Movement, the Congress expected to use the anger against the Gandhi arrest to start a nation-wide agitation.

After Patel's arrest, feelings ran so high that they refused to cool down, especially because rumours of Gandhi's arrest were everywhere. It was even suggested that he would be sent to jail in Rangoon.

As a result, Sabarmati Ashram was overrun with visitors on Monday 10 March, two days before the march. Normal attendance at evening prayer meetings was generally under five hundred. Suddenly there were two thousand people or more. The prayer ground was too small for them; the congregation had to move to the river bed. Gandhi wasn't exactly happy that so many of them had come; in fact he had serious reservations, which he expressed sternly, in the plain-speaking way which was so characteristic of him:

> I am glad that you have been coming to our prayer in such large numbers, and generally I would say, "May your tribe increase." But I must utter a few words of warning. If it is mere curiosity

that draws you here, you had better not come at all. If it is the prayer that attracts you, you are quite welcome, but in that case this sudden inroad cannot be accounted for. But I presume you come both for the prayer and for understanding the significance of the campaign I am about to launch.

As for the prayer, I assure you that mere utterance parrot-wise of the name of God is of no avail. All your trouble in coming this long distance from the town would be wasted and the quiet of our prayers would be disturbed. If therefore your desire to take part in the prayer is genuine, you must be prepared to fulfil a condition which alone can prove your bona fides, and that is that you come here dressed in khadi. You may or may not admit the many claims made on behalf of khadi, but one thing is now practically universally admitted, that khadi unites the wearer to the poorest of the land. And I may tell you that but for the progress that khadi has made in recent years, I should not have been able to launch this campaign. It is the spread of khadi that infuses in me the hope that the message of non-violence has spread with khadi. A believer in violence may wear khadi, but he would do so in order to exploit it.

As for the other desire that also prompts you to come here, you know that the march begins on Wednesday morning. Every one is on the tiptoe of expectation, and before anything has happened the thing has attracted worldwide attention. Now I should like to analyse the thing for you and to implore you to appreciate its implications. Though the battle is to begin in a couple of days, how is it that you can come here quite fearlessly? I do not think any one of you would be here if you had to face rifleshots or bombs. But you have no fear of rifleshots or bombs. Why? Supposing I had announced that I was going to launch a violent campaign (not necessarily with men armed with rifles, but even with sticks or stones), do you think the government would have left me free until now? Can you show me an example in history (be it England, America or Russia), where the State has tolerated violent defiance of authority for a single day? But here you know that the government is puzzled and perplexed. And you have come here, because you have been familiarized by now with the idea of seeking voluntary imprisonment.

Then I would ask you to proceed, a step further. Supposing ten men in each of the seven lakh villages in India come forward to manufacture salt and to disobey the Salt Act, what do you think can this government do? Even the worst autocrat you can imagine would not dare to blow regiments of peaceful civil resisters out of a canon's mouth. If only you will bestir yourselves just a little, I assure you we should be able to tire this government out in a very short time. I want you therefore to understand the meaning of this struggle and to do your part in it. If it is only curiosity that moves you to walk this long distance, you had better not waste your time and mine. If you come here to bless us and our Movement, the blessings must take some concrete shape. I don't want any money from you. I am hoping that it may be possible to fight this battle with the least possible money. At the time of Kheda Satyagraha in 1918 I had to refuse several offers for raising contributions. In Bardoli an appeal was made and there was a spontaneous response, but much of the money was saved and is now being utilized for constructive work. So I do not want you to contribute any money just now. That you will do unasked when our suffering has reached that stage which cannot but compel your sympathy. But I want you to take your courage in both hands and contribute in men toward the struggle which promises to be fierce and prolonged. I certainly expect the city of Ahmedabad, the Ahmedabad of Vallabhbhai who is already in jail, to furnish an unlimited supply of volunteers to keep the stream unbroken, in case batch after batch happens to be arrested and marched to jail. That is the least I expect of you. May God give you the strength to rise to the occasion.

The atmosphere at the ashram can be imagined: a day before the march was to begin, it was nearing a state of frenzy. Not because of Gandhi and the marchers; it was because, as the prayer

meeting showed, the outside world had reached a kind of fevered excitement. The fever had particularly affected the press and radio, and reporters and photographers besieged Sabarmati Ashram. The man in the eye of the storm remained calm. He carried on with his normal routine, but did fit in interviews with the media.

All this while, the selection of the Playing XI, so to speak, had been in progress. The selection committee consisted of Mahatma Gandhi (Chairman), Mohandas Karamchand Gandhi, M.K. Gandhi and Mr Gandhi. In other words, it was a one-man committee, with no court of appeal and no arbitration clause.

As we have seen, for very rational reasons, Gandhi had decided that the marchers would only be from Sabarmati Ashram. And in the ashram, it was he who knew everyone: he was father, father-confessor, mentor, guru, teacher ... all rolled into one. There was no aspect of an ashramite's life that Gandhi did not have full knowledge of.

He began by eliminating women for a strategic reason already outlined. He also ruled out foreigners, even those committed to the nationalist cause, because this was an Indian freedom struggle and though Gandhi might have used Reginald Reynolds as a courier of an important letter, he wasn't about to use him as a marcher. Reynolds, no doubt, was devastated.

There were other factors in the elimination process. Some ashramites were, in Gandhi's opinion, too young. There were others who were, in any opinion, too old. One such was Imam Saheb Abdul Qadir Bawazir, an old ashramite and 'co-worker'. What good were old loyalties, old associations, old beliefs in principles, if the legs too were old? The Imam Saheb would have been a propaganda coup, but the propaganda would be short-lived.

Gandhi ruled out some others on equally pragmatic grounds: there was a request for enlistment from a father, two sons and son-in-law. Gandhi accepted all of them except the older son: he didn't want an entire family wiped out if there was a violent reaction from the police. Others kept back were those who had an important role to play in the ashram.

Almost all the marchers were young men, which was to be expected given the long distance to be covered on foot. In fact, most of them were between twenty and twenty-five years old. The exceptions was Vithal Liladham Thakkar, a 16-year-old student at the ashram school, and at the other end of the scale, a 60-year-old man who also lived at the ashram. He answered to the name of Mohandas Karamchand Gandhi.

The indefatigable Mahadev Desai, whose articles often appeared under the initials M.D., anticipating that the media would want the names and some details of the marchers, published the following list:

> The first batch of satyagrahis number seventy-nine. All these men are practically the inmates of the ashram, and therefore the provincial boundary has no significance. It will interest the reader to know that a provincial list was prepared for the first time for the sake of these notes. At the ashram we often do not know the province or the caste or religion to which a man belongs. The list according to provinces is quite interesting. Here it is: 32 belong to Gujarat, 13 belong to Maharashtra, 7 belong to the UP, 6 to Cutch, 3 belong to the Punjab, 1 to Sindh, 4 to Kerala, 3 to Rajputana, 1 to Andhra, 1 to Karnataka, 2 to Bombay, 1 to Tamilnadu, 1 to Bihar, 1 to Bengal, 1 to Utkal, 1 to Fiji (originally of UP but born in Fiji), 1 to Nepal.

Desai did a further analysis to forestall the inevitable questions:

> Considered according to communities broadly, there were 2 Musalmans, 1 Christian and the rest are Hindus (2 representing the so-called "untouchables"). The work they are doing in the ashram shows 99 teachers, 25 khadi students and the rest are either in charge of the various departments of the ashram or on

the office staff. From the point of view of academic qualifications there are 12 graduates; 7 of Bombay University, 3 of the Vidyapith, and 2 of foreign universities ... No doubt the reader is familiar with some of the names. But he need not know anything beyond the fact that they are India's soldiers of peace. (It was later clarified that there were not two but four "untouchables" in the list.)

It wasn't accidental that though Sabarmati Ashram was in Ahmedabad in the heart of Gujarat, the group of marchers represented the whole of India.

A couple of things stand out in the list: it has two 'foreigners'. The Fiji citizen's presence is explained by Desai 'originally of UP but born in Fiji'. The Nepali had an interesting history. Kharaj Bahadur Singh was a convicted killer who had served a sentence for murder. But he had reformed, he said, and Gandhi believed him. One of the two foreign university graduates was Haridas Mazumdar, who became 'Haribhai' in the ashram. He had studied in the United States of America and had become an academic, but his nationalist fervour had brought him to Sabarmati Ashram as a guest. Although he was thoroughly Westernized (one story talks about his futile shopping forays in Ahmedabad for knickerbockers) he tried to live the life of an ashramite, though not always successfully. No one expected him to be selected, not even he. Gandhi's reasons for choosing him may have been to include a 'foreign-returned' scholar, and one who could be relied upon to go back and talk about the march and Satyagraha to an American audience.

There is also the story of the prominent Muslim leader Abbas Tyabji. When he saw Desai's list, he was overjoyed that 'Abbas' was on it, but also a little surprised: he was, after all, seventy-six and hadn't expected to be included. It turned out that there was a younger Abbas, a 22-year-old teacher in the ashram's technical school, who was supposed to go. Gandhi made amends by asking Tyabji to go in a car and receive the marchers at designated halts. More than that, he gave Abbas Tyabji a very real honour by designating him as his replacement when he was arrested, the role Sardar Patel was to have taken on.

Even at this juncture of high excitement – or, perhaps, particularly at this point – Gandhi felt the need to spell out his thoughts and his vision. This was in keeping with his practice of complete candour and transparency: a satyagrahi never tells anything but the truth; more than that, he never hides the truth. He, therefore, asked Mahadev Desai to print another Q&A session in *Young India*, one which would spell out Gandhi's plan when Swaraj was obtained, the pros and cons of following a strategy of boycotting foreign goods versus Civil Disobedience and the risk of violence even in Satyagraha. In a way he was addressing people's concerns on the eve of an historic event.

The Q&A was titled 'Talks before the Trek':

Q. What sort of Government do you want?
A. I want a Government that would obey and carry out the wishes of the people.
Q. You want a democracy?
A. I am not interested in words, and I never worry myself about the form of government.
Q. But don't you mind methods?
A. I do mind them very much indeed, but I don't mind the form.
Q. Then you would not mind a monarchy?
A. I said form and machinery do not much matter to me.
Q. Well then, tell me what form your democracy will take?
A. I do not know; the method alone interests me, and by method I mean the agency through which the wishes of the people are reached. There are only two methods: one is that of fraud and force; the other is that of non-violence and truth. Force always includes fraud, non-violence always excludes it.
Q. Can't you have fraud with non-violence?

A. No. Impossible. Fraud itself is a species of violence.

Q. Well, I have seen fraud associated with non-violence. China is reputed to be one of the most peaceful countries in the world, and if I were to tell you about the frauds there, you would be shocked.

A. I repeat, words do not capture my imagination. As a people, the Chinese are one of the most peaceful in the world, but that peace cannot be real and voluntary if there is fraud in it. If I harbour ill will in my breast but do not express in its acts, I am still violent. By non-violence of peace I mean the peace which comes through inner strength. If I have that peace, that non-violence, I won't have any hate in me. Violence does not necessarily mean bodily harm. What I want to impress on every one is that I do not want India to reach her goal through questionable means. Whether that is possible or not is another question. It is sufficient for my present purpose if the person who thinks out the plan and leads the people is absolutely above-board and has non-violence and truth in him. Non-violence works organically, not mechanically. It was for that reason that I asked for unrestricted authority from the Working Committee of the Congress to work out my plan of non-violence.

BOYCOTT OF FOREIGN GOODS v. C.D.

Q. Don't you think, Gandhiji, boycott of foreign goods would be more potent than Civil Disobedience?

A. Years ago I heard that futile cry and I replaced it by one for the boycott of foreign cloth. It made some impression, but that of boycott of foreign goods made none at all.

Q. My impression is that in Bengal boycott of British goods was tried with success, but no other province took it up.

A. No. It fizzled out. The mills in Ahmedabad and Bombay defrauded the nation by sending spurious cloth; and when genuine mill cloth was sent, exorbitant prices were charged.

Q. That is what I mean to say. The thing was not tried seriously.
A. If it was not, it means that the people did not want to. So far as I am concerned, I never believed in it and so I could never back it.
Q. But would it not be easier to handle foreign cloth boycott than Civil Disobedience?
A. No. It is far more difficult. In one case you want the cooperation of 300 million people. In the other, even if you have an army of ten thousand defiant men and women, your work is done.
Q. Why? They can be all clapped in jail and nothing more will happen thereafter.
A. Let them try the experiment. They would have to hang these men before they could dismiss them from their minds. If these men are trusty and true their very presence will worry the government unto death.
Q. Will they worry government even in prison?
A. For one thing they can't keep them there for any length of time. The fact is that we never had even five thousand full civil resisters in 1921. Every political prisoner is not necessarily a civil resister.

THE RISK OF VIOLENCE

Q. Will not your Movement lead to violence?
A. It may, though I am trying my best to prevent any outbreak of violence. Today there is greater risk of violence, in the absence of any safety-valve in the shape of a Movement of non-violence like the one I am contemplating.
Q. Yes, I have heard you say that you are launching this campaign for the very purpose of stopping violence.
A. It is one argument, but that is not the most conclusive argument. The other and most conclusive argument for me is that if non-violence has to prove its worth, it must prove its worth today. It must cease to be the passive or even

impotent instrument that it has come to be looked upon in certain quarters. And when it is exercised in the most effective way, it must act in spite of the most fatal outward obstructions. In fact non-violence by its very nature must neutralize all outward obstruction. On the contrary, inward obstacles in the shape of fraud, hatred, and ill will would be fatal to the Movement. Up to now I used to say, "Let me get control over the forces of violence." It is growing upon me now that it is only by setting the force of non-violence in motion that I can get those elements under control.

But I hear people say, "History will have to repeat itself in India." Let it repeat itself, if it must. I for one must not postpone the Movement unless I am to be guilty of the charge of cowardice. I must fight unto death the system based on violence and thus bring under control the force of political violence. When real organic non-violence is set to work, the masses also will react manfully.

A MIRACLE

Q. But after you are removed the Movement will no longer be in your control?

A. In South Africa the movement was not in my control during the latter part of it, when it gained considerable momentum without any action on my part. Thousands joined the movement instinctively. I had not even seen the faces of them, much less known them. They joined because they felt that they must. They had possibly only heard my name, but they saw in the twinkling of an eye that it was a movement for their liberation; they knew that there was a man prepared to fight the £3 tax and they took the plunge. And against what odds? Their mines were converted into jails; the men who oppressed them day and night were appointed warders over them. They knew that there would be hell let loose on them. And yet they did not waver or falter. It was a perfect miracle.

THE OPPORTUNITY OF A LIFETIME

Q. But would not the Movement add to the already numerous divisions existing in the country?

A. I have no such fear. The forces of disunion can be kept under control, even as the forces of violence. You must say that there is fear elsewhere. The party of violence may not respond to my advances and the masses might behave unthinkingly. I am an optimist and have an abiding faith in human nature. The party of violence will give me fair play and the masses will act rightly by instinct. It is possible that I may be living in a fool's paradise. But no general can possibly provide for all contingencies. For me it is the opportunity of a lifetime. The Movement is none of my seeking. Almost in spite of myself I was irresistibly drawn to Calcutta. I entered into a compromise to which I was driven. The period of two years I changed to one, simply because it did not involve any moral principle. In Lahore I had to conceive and frame practically every resolution. There I saw the force of violence and non-violence in full play, acting side by side; and I found that non-violence ultimately triumphed over violence.

HOW IS THE TIME RIPE NOW?

Q. You said some time ago that the time was not ripe for Civil Disobedience. What has happened between that time and today that has helped you to alter your view?

A. I am quite positive that it is fully ripe. The reason I will tell you. Nothing has happened externally, but the internal conflict in me, which was the only barrier, has ceased; and I am absolutely certain now that the campaign had been long overdue. I might have started it long before this.

Q. And what was that internal conflict?

A. You know I have always been guided solely by my attitude

towards non-violence, but I did not know then how to translate that attitude into action in the face of growing violence. But now I see as clearly as daylight that, pursuing the course that I have adopted, I minimize the risk I am taking.

Q. Are you sure that the salt campaign will lead you to jail?
A. I have not a shadow of doubt that it will. How long exactly it will take is more than I can say, but I feel that it will be much sooner than most people would be inclined to think. I expect a crisis to be soon reached which would lead to a proper conference – not a Round Table conference, but a Square Table one where everybody attending it would know his bearings. The exact lineaments of that Conference I cannot at present depict, but it will be a conference between equals to lay their heads together to devise ways and means for the establishment of an Independent Constitution in India.

Nothing more needed to be done. Nothing more could be done: the march was to begin the next day, an event Gandhi and the Congress leadership expected to lead to a complete Civil Disobedience Movement and ultimately to Purna Swaraj, Complete Independence. But before that they expected the other side to act to prevent the inevitable, even to retaliate with the only weapons they knew. Satyagraha meant that your opponent knew your moves before you knew his, but it also meant that you would disarm him with this very lack of knowledge. Would Gandhi be in jail? Certainly no one knew. Would the other marchers be arrested? Probably. When? No one knew. Would there be violence? Possibly. When? No one knew. The marchers retired to their quarters on 11 March with these troubling thoughts. But they weren't the only ones on early call.

ABOVE:
Mahatma Gandhi and satyagrahis on the way to Dandi on the eve of the Salt March in March 1930.
Photo courtesy: Gandhi Smarak Sangrahalaya Smriti

RIGHT:
Gandhi marching with other satyagrahis. The masses always identified with Gandhi's political approach and thronged to his call.
Photo courtesy: Roli Collection

INDIAN POSTS AND TELEGRAPHS DEPARTMENT

Received here at 11 H. 12 M.

O. IG. 9/35

AHMEDABAD SABARMATI   12   18

JAWAHARLAL NEHRU ALLAHABAD

BAPU MARCHED OFF THIS MORNING MIDST WONDERFUL SCENES THOUSANDS FOLLOWING FOR MILES

—MIRA—

This form must accompany any inquiry respecting this Telegram.

ABOVE:
Telegram from Mira Ben to Nehru.
Photo courtesy: Gandhi Smarak Sangrahalaya Smriti

RIGHT:
Wounded satyagrahi being carried on a stretcher during the lathi charge at Dharasana. *Photo courtesy*: Gandhi Smarak Sangrahalaya Smriti

FACING PAGE (TOP):
The end of a long day of marching.
Gandhi with young satyagrahis.
Photo courtesy: Gandhi Smarak Sangrahalaya Smriti

FACING PAGE (BOTTOM):
Telegram from Gandhi to Jawaharlal Nehru.
Photo courtesy: Nehru Memorial Library

C. 3.

INDIAN POSTS AND TELEGRAPHS DEPARTMENT.

HL 8/55   Received here at ___ H. ___ M.   No.
AHMEDABAD SADARMATI
         ∧⁶   18

£/ JAWAHARLAL NEHRU ALLAHABAD

LETTER HANDED PRESS FOR PUBLICATION MARCHING EARLY MORNING 12TH WITH

SIXTY COMPANIONS.

GANDHI.

This form must accompany any inquiry respecting this Telegram.

ABOVE:
Gandhi with Sarojini Naidi at Dandi.

RIGHT:
Gandhi and Kasturba with satyagrahis taking a rest during the march.

FACING PAGE:
Gandhi taking the historic first fistful of sand and salt at Dandi, April 1930. Mithin Behn Patil is with him.
Photos courtesy: Gandhi Smarak Sangrahalaya Smriti

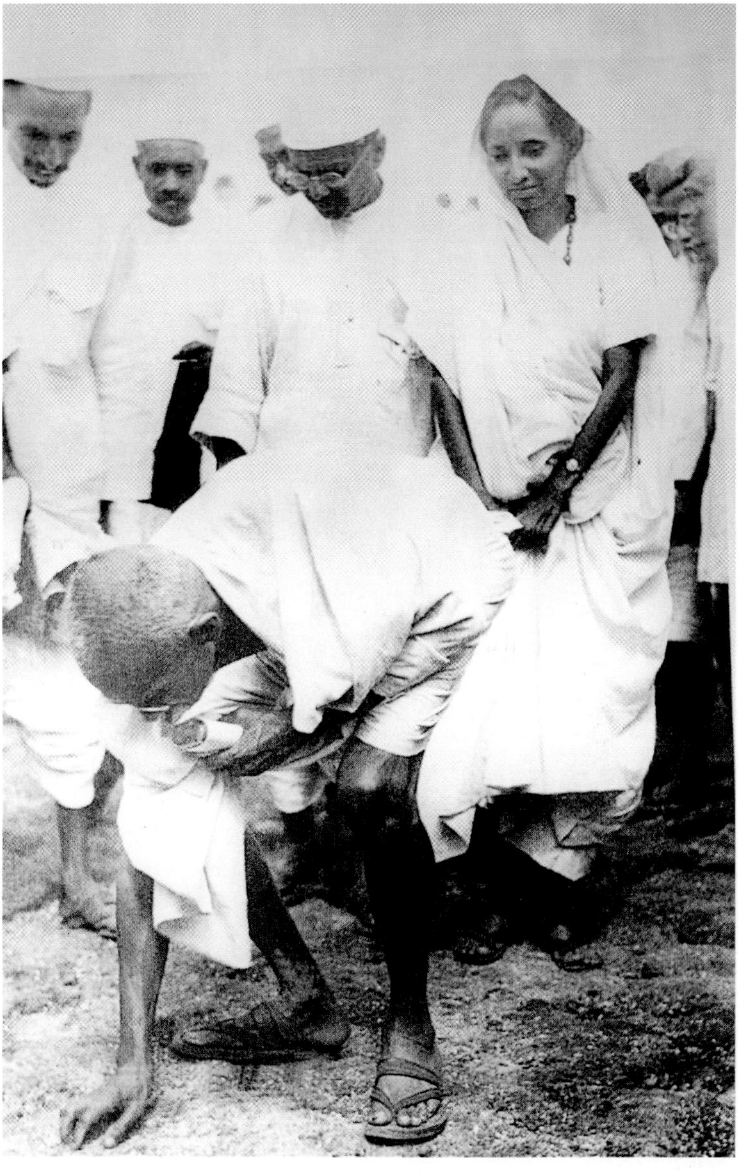

ABOVE:
Dr Annie Besant, (centre) with a group of Indians protesting against the Salt Tax. To her left is the Hon. Mr Rangacharia, Member of the Legislative Assembly.

FACING PAGE (TOP):
Young Nationalist supporters of Mahatma Gandhi break salt laws by filling containers with sea water, at Bombay beach, 1930.

FACING PAGE (BOTTOM):
Manufacturing contraband salt at the beach in Madras.
Photos courtesy: Hulton Getty Archive

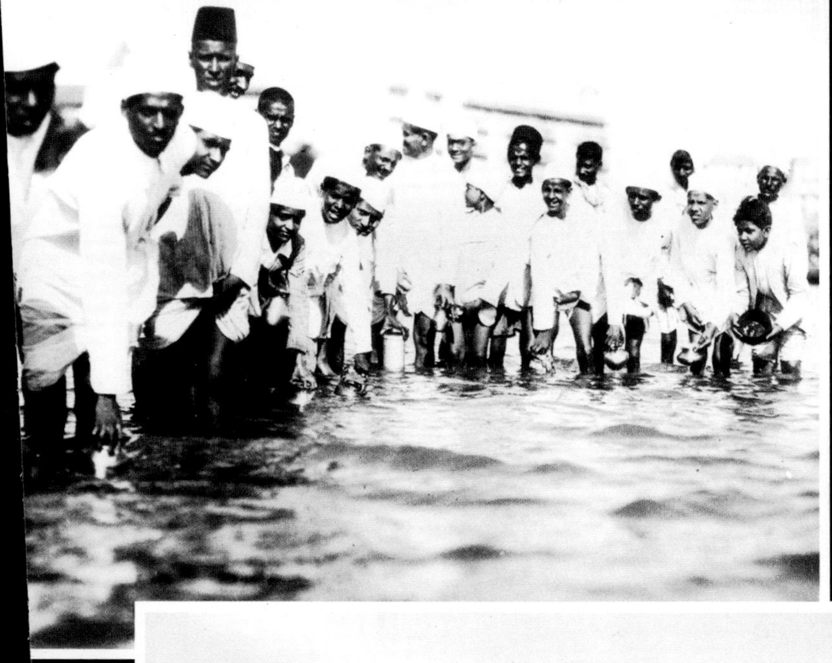

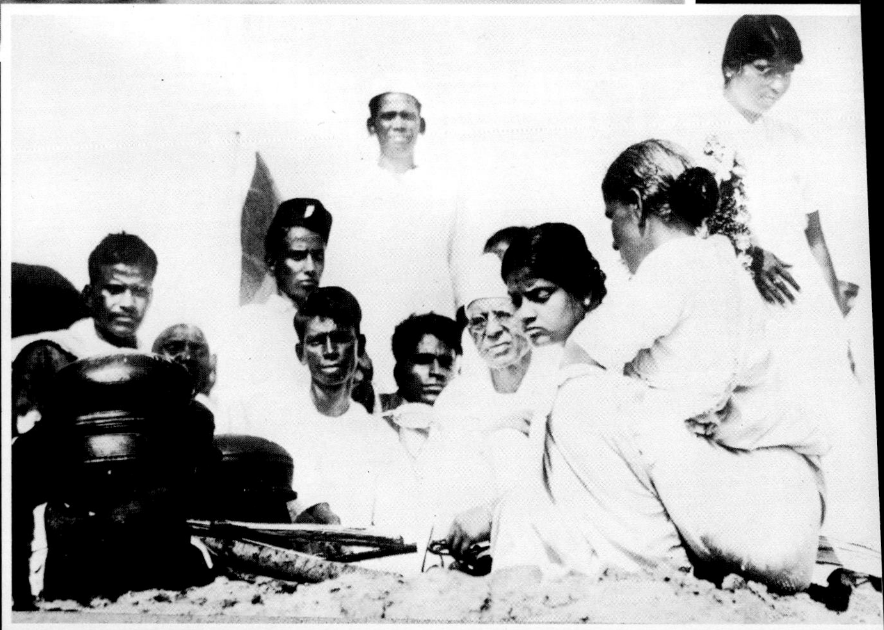

Mahatma Gandhi's message during the Salt Satyagraha, 1930.
Photo courtesy: Gandhi Smarak Sangrahalaya Smriti

THE MARCH TO DANDI

The eve of the march became a night without end. Crowds had begun to gather through the day so that at the time of the usual evening prayers, there were ten thousand people milling around, far too many for the ashram and its prayer ground. Once again everyone was made to assemble on the dry banks of the Sabarmati.

After the prayers, Gandhi spoke to the assembled crowd.

In all probability this will be my last speech to you. Even if the government allows me to march tomorrow morning, this will be my last speech on the sacred banks of the Sabarmati. Possibly these may be the last words of my life here.

Today I shall confine myself to what you should do after I and my companions are arrested. The programme of the march to Jalalpur must be fulfilled as originally settled. The enlistment of volunteers for this purpose should be confined to Gujarat. From what I have seen and heard during the last fortnight I am inclined to believe that the stream of civil resisters will flow unbroken.

But let there be not a semblance of breach of peace even after all of us have been arrested. We have resolved to utilize all our resources in the pursuit of an exclusively non-violent struggle. Let no one commit wrong in anger. This is my hope and prayer. I wish these words of mine reached every nook and corner of the land. My task shall be done if I perish and so do my comrades. It will then be to the Working Committee of the Congress to show you the way and it will be up to you to follow its lead. That is the only meaning of the Working Committee's resolution. The reins of the Movement will still remain in the hands of those of my associates who believe in non-violence as an article of faith. Of course, the Congress will be free to chalk out what course of action commends itself to it. So long as I have not reached Jalalpur, let nothing be done in contravention to the authority vested in me by the Congress. But once I am arrested, the whole general responsibility shifts to the Congress. No one who believes in non-violence, as a creed, need therefore sit still. My compact with the Congress ends as soon as I am arrested. In that case there should be no slackness in the enrolment of volunteers. Wherever possible, Civil Disobedience of salt laws should be started. These laws can be violated in three ways. It is an offence to manufacture salt wherever there are facilities for doing so. The possession and sale of contraband salt (which includes natural salt or salt earth) is also an offence. The purchasers of such salt will be equally guilty. To carry away the natural salt deposits on the seashore is likewise a violation of law. So is the hawking of such salt. In short, you may choose any one or all of these devices to break the salt monopoly.

We are, however, not to be content with this alone. Wherever there are Congress Committees, wherever there is no ban by the Congress and wherever the local workers have self-confidence, other suitable measures may be adopted. I prescribe only one condition, viz., let our pledge of truth and non-violence as the only means for the attainment of Swarajya be faithfully kept. For the rest, every one has a free hand. But that does not give a licence to all and sundry to carry on their individual responsibility. Wherever there are local leaders, their orders should be obeyed by the people. Where there are no leaders and only a handful of men have faith in the programme, they may do what they can, if they have enough self-confidence. They have a right, nay it is their duty, to do so. The history of the world is full of instances of men who rose to leadership, by sheer force of self-confidence, bravery and tenacity. We too, if we sincerely aspire to Swarajya and are impatient to attain it, should have similar self-confidence. Our ranks will swell and our hearts strengthen as the number of our arrests by government increases.

Let nobody assume that after I am arrested there will be no one left to guide them. It is not I but Pandit Jawaharlal who is your guide. He has the capacity to lead. Though the fact is that those who have learnt the lesson of fearlessness and self-effacement need no leader, but if we lack these virtues, not even Jawaharlal will be able to produce them in us.

Much can be done in other ways beside these. Liquor and foreign cloth shops can be picketed. We can refuse to pay taxes if we have the requisite strength. The lawyers can give up practice. The public can boycott the courts by refraining from litigation. Government servants can resign their posts. In the midst of the despair reigning all round people quake with fear of losing employment. Such men are unfit for Swarajya. But why this despair? The number of government servants in the country does not exceed a few hundred thousand. What about the rest? Where are they to go? Even free India will not be able to accommodate a greater number of public servants. A Collector then will not need the number of servants he has got today. He will be his own servant. How can a poor country like India afford to provide a Collector with separate servants for performing the duties of carrying his papers, sweeping, cooking, latrine cleaning and letter carrying? Our starving millions can by no means afford this enormous expenditure. If, therefore, we are sensible enough, let us bid goodbye to government employment, no matter if it is the post of a judge or a peon. It may be difficult for a judge to leave his job, but where is the difficulty in the case of a peon? He can earn his bread everywhere by honest manual labour. This is the easiest solution of the problem of freedom: Let all who are cooperating with the government in one way or another, be it by paying taxes, keeping titles, or sending children to official schools, etc., withdraw their cooperation in all or as many ways as possible. One can devise other methods too of non-cooperating with the government. And then there are women who can stand shoulder to shoulder with men in this struggle.

You may take it as my will. It was the only message that I desired to impart to you before starting on the march or for the jail. I wish there to be no suspension or abandonment of the war that commences tomorrow morning, or earlier if I am arrested before that time. I shall eagerly await the news that ten batches are ready as soon as my batch is arrested. I believe there are men in India to complete the work begun by me today. I have faith in the righteousness of our cause and the purity of our weapons. And where the means are clean, there God is undoubtedly present with His blessings. And where these three combine, there defeat is an impossibility. A satyagrahi, whether free or incarcerated, is ever victorious. He is vanquished only when he forsakes truth and non-violence and turns a deaf ear to the Inner Voice. If, therefore, there is such a thing as defeat for even a satyagrahi, he alone is the cause of it. God bless you all and keep off all obstacles from the path in the struggle that begins tomorrow. Let this be our prayer.

That should have been that. The crowd should have wished Gandhi and his marchers 'Good Luck and Godspeed' and melted away to their own homes. They didn't. In fact, no one wanted to leave.

In this large crowd, rumours swirled around serially: one would do the rounds (a car was on its way to arrest Gandhi and put him in a jail in Burma), soon to be followed by another (no it wasn't a car; the British had arranged a special train). A police lorry (or lorries) were reliably learnt to be scheduled to arrive at the ashram at 11 o'clock, two hours after its official closing time... The rumours were fuelled by the occasional car driving up from the direction of Sabarmati; on the dirt road, the gravel made silence impossible even for a purring engine, and in any case, the dust thrown up by the wheels rose as an unmistakable signal in the air.

'Here they come,' went the shouts every time a car drove up, but no car seemed to be the right one. Yet the vigil continued, with the crowd, tense but peaceful, singing bhajans while they waited, and occasionally shouting 'Mahatma Gandhi ki jai'.

The Mahatma himself carried on with his routine as if this was a normal day and as if there were no crowds at all waiting outside making all that noise. In a small departure from his usual routine, he hadn't gone to sleep even at ten. That's when he sat down and wrote to Jawaharlal Nehru. That duty being done he lay down on his bed, and before anyone knew it, was fast asleep. The crowd had disappeared. Certainly not physically, but most definitely from Gandhi's mind.

Two khadi bags had been packed for him. One contained essentials, the other, bare essentials. The idea was to jettison the first if he was arrested and only take the second to jail. We have seen the photographs countless times: the bags, one on either side, are on Gandhi's shoulders as he strides forward, staff in hand. In any case, for Gandhi, travelling light wasn't a new experience: he lived a life with no material possessions.

Except for Gandhi, no one really slept, so the usual 4 a.m. wake up bell was somewhat redundant. All the people who comprised the previous night's crowd hadn't gone away. They had nowhere to camp but in the open. So they were certainly awake, and they were a bit cold, in spite of the bonfires many had lit. They had already filled the river bank prayer space by the time Gandhi came to speak to them.

He spoke, expectedly, of the march which was to begin in just over two hours. His tone was sombre in a speech directed mainly at the marchers. To them his speech was part inspirational, part warning of dangers ahead. Shape up or ship out, even NOW, he seemed to say:

> Those inmates of the ashram who have any dependents will not be able to draw money from the ashram for them. None should join in the struggle with that hope. This fight is no public show; it is a final struggle – a life and death struggle. If there are disturbances, we may even have to die at the hands of our own people... We have constituted ourselves the custodians of Hindu-Muslim unity. We hope to become the representatives of the poorest of the poor, the lowest of the low, the weakest of the weak. If we do not have the strength for this, we should not join the struggle... I ask you to return here only as dead men or as winners of Swaraj... Even if the ashram is on fire, we will not return. Only those may join, who have no special duty to their relatives... We are entering upon a life-and-death struggle, a holy war; we are performing an all-embracing sacrifice in which we wish to offer ourselves as oblation. If you are incapable, the shame will be mine, not yours. You too have the strength that God has given me.

Speech over, Gandhi lay down again, unaffected by the tumult around him. At ten past six he was out and ready (he had asked all marchers to be present at 6.20 for 6.30 departure). All seventy-eight got in line, two abreast, in the order already indicated. Gandhi, of course, was at the head of the column.

Everyone was in white khadi, most in dhotis, some wearing kurtas, others a chadar with their dhoti. Each marcher carried a bag with everything he needed for the march: a change of clothes, notebooks and pen, a bedding roll, a tin thali and a mug and prayer beads.

What is surprising is that the group of marchers carried no placards or identifying tags. Perhaps the latter were superfluous, but the absence of the former was unexpected. It is possible that

Gandhi, who relied so much on the spoken and the written word, wanted that in the Dandi March, actions should speak louder: that the fact of the walking itself, and the picking up of salt later would convey the message. There must have been the simple practical consideration as well: placards are cumbersome to carry and in such a long march, the lighter the load, the better.

Thomas Webber narrates a story regarding badges. Apparently Jawaharlal Nehru thought of it as a good idea and actually distributed them to the marchers; Gandhi's badge he playfully pinned surreptitiously on his chadar. When Gandhi saw it, he didn't take it off, but made sure it got into the folds so that no one could see it.

All the ashram's women, and the men who were staying behind, formed the farewell party. Kasturba began by garlanding Gandhi, not with flowers but with khadi thread. Kumkum was smeared on his forehead. This informal ceremony was being carried out throughout the length of the column.

Gandhi kept looking at his watch, the large fob he always carried. At exactly 6.30 a.m., he took the first step, with the column joining in. All around them rupee coins were showered, coconuts broken and slogans shouted. A ragtag band, which had collected unbidden, broke into music. Absurdly, they began by playing the notes of *God Save the King*. Someone noticed the incongruity, and stopped them.

The crowd began to shout 'Mahatma Gandhi ki jai' while the marchers sang Gandhi's favourite hymn, *Vaishnva Jana To*, accompanied, discordantly, by the blowing of conch shells from the crowd. It was cacophony, in every sense of the word. If the marchers sang a bhajan, it was probably to keep their courage up. Of all the wild stories going around, the wildest were of a battery of machine guns waiting to mow down everyone. Or of a plane waiting to bomb them all, as soon as the marchers were in a clearing. These rumours may sound ridiculous in retrospect, but the Jalianwala Bagh massacre was fresh in everyone's mind, and *that* was brutal, unprovoked mass murder of a peaceful and unarmed crowd. Just like the marchers.

The mass media of the day was present in large numbers: reporters, radio journalists and documentary news film crews, many in jeeps, some in hired trucks. They added to the din, as they shouted instructions and entreaties, recorded commentaries and orchestrated reactions from all over the country. 'Like the historic march of Ram Chandra to Lanka,' is the way Motilal Nehru saw it. P.C. Ray likened it to the 'Exodus of Israelites under Moses'. Jawaharlal Nehru was more poetic:

> Today the pilgrim marches onward on his long trek. Staff in hand he goes along the dusty roads of Gujarat, clear-eyed and firm of step, with his faithful band trudging along behind him. Many a journey he has undertaken in the past, many a weary road traversed. But longer than any that have gone before is this last journey of his, and many are the obstacles in his way. But the fire of great resolve is in him and surpassing love of his miserable countrymen. And love of truth that scorches and love of freedom that inspires. And none that passes him can escape the spell, and men of common clay feel the spark of life. It is a long journey, for the goal is the independence of India and the ending of the exploitation of her millions.

Whatever the mode of expression, it was obvious that the march was being seen as one on an epic scale, an event that would not just go down in history, but would *change* the course of history.

In complete contrast to the tumult accompanying the marchers, there was a sense of silent gloom at Sabarmati Ashram. The place felt physically empty. More than that, its moving force (and guiding light) was no longer there. Even worse, aligned to the feeling of pride and expectation of what was to be accomplished, was a feeling of deep foreboding, made worse by some words in Gandhi's speech the previous night; 'It is a final struggle – a life and death struggle... We may even have to die at the hands of our own people.' Then the chilling finale, '(We) return here only as dead men or as winners of Swaraj.' No wonder everyone who had stayed behind (or as generally was the case, was left behind), was desolated.

This feeling of foreboding was based on apprehensions of how the British would act. It was clear that the worst was expected. This was, in a way, given the confirmation everyone was looking for in the quick, and preemptive, arrest of Vallabhbhai Patel.

But if the British had a game plan, no one knew it. Not even, it seems, the British. As we have seen, Patel's arrest was a case of someone jumping the gun. Its fallout was there for all to see: it had given the Dandi March just the fillip it needed. It would have been easy enough to arrest Gandhi, but what repercussions would that have? Lord Irwin preferred to wait and watch.

As in the case of Patel, there was a definite hierarchy of concern about the Dandi March: the Government of India and other government bodies at the centre were of the opinion that, given time, it would play itself out; the Government of Bombay, under whose jurisdiction the march was taking place, viewed it with growing concern; district level Collectors and other officials were positively alarmed and called for quick action.

This hierarchy of concern can be seen from the various pronouncements and missives which went to and fro within the British administration. A member of the Central Board of Revenue, George Richard Fredrick Totteham, for example, called the march 'Mr Gandhi's somewhat fantastic project'. Lord Irwin in a telegram to the secretary of state for India in London, twelve days into the march, said 'So far as Gujarat is concerned, local effect is considerable, though possibly transient. As regards the rest of India, the present effect seems to be comparatively small, and the Government of India are inclined to the view therefore that Gandhi should not be arrested unless matters so develop that there is no reasonable alternative or unless Government of Bombay feel the requirements of the local situation, urgently demand such action.' In a more informal message to the Archbishop of Canterbury, Irwin said, 'I discussed it all with Sykes and Hailey a few days ago and we are all agreed that it was right to avoid arresting him as long as one could on his silly salt stunt...' In short, Irwin felt that if no action was taken, Gandhi would 'destroy himself by ridicule'.

In complete contrast are the words of Fredrick Sykes, the Governor of Bombay:

> Both Master, the Collector of Kaira, and Garrett, the Commissioner of the Northern Division, were anxious to put an end to the march altogether. From the outset they thought that Gandhi was being given too much rope and that the policy of allowing him to make himself ridiculous was based on a fallacy.
>
> Though I personally was uneasy about the propaganda effect of the march, it was essential that we should act in conformity with the wishes of the Central Government. I accordingly issued confidential orders to the officials through whose districts Gandhi and his followers were to pass. I pointed out that in the opinion of the Government of India, the ordinary law did not permit any interference with the march, provided that it was undertaken in a peaceful and orderly manner. Even though salt-making was illegal, the offence was so trifling as to be unworthy of serious notice. The premature arrest of Gandhi would merely give him the spectacular martyrdom on which he was courting, and would, in addition, probably lead to an outbreak of violence among his followers. The maximum term of imprisonment to which Gandhi would be liable, if arrested during his march, would be much less than if we held our hand until an offence was actually committed.

(Incidentally, Section 117 of the Indian Penal Code, under which Gandhi could have been arrested, was a bailable offence. So, in theory, he could have got bail and finished the march, thus striking a double blow to the British.)

Local administrations, in the meantime, were doing whatever they could at their level: one of the first steps was to withdraw the censor certificates for a film on the march. Mahadev Desai, sitting at Sabarmati Ashram, wrote:

> The Governments of Bombay, the UP and the Punjab have thought it fit to declare as uncertified three films of "Mahatma Gandhi's March for Freedom on the 12th of March 1930." It will be remembered that the Bombay Board of Film Censors had certified them as unobjectionable, but the Governments of Bombay, the UP and the Punjab had obviously reasons which were beyond the ken of the Board of Censors. If these films were so much explosive stuff, one wonders why the actual march, which must be a hundred times more explosive, should be allowed to continue from day to day. But it is no use trying to dissect the strategy of government. It may be foolish, it may be deceptive, it may be anything; the only strategy of the satyagrahi is truthful and non-violent action in broad daylight. It may however be necessary to tell everyone concerned that India's battle for freedom does not depend in the least on cinema films. The struggle which was auspiciously begun on the 6th of April 1919, and which "puzzled and perplexed" Lords Reading and Lloyd did not depend on cinema films and had not to be interrupted for want of cinema films.
>
> And then there were no cinema films in the days when Rama marched on Lanka, nor in the days of the memorable war between the Pandavas and the Kauravas. And yet there is not an Indian but knows the story of the *Ramayana* and the *Mahabharata*, and there are thousands today whose memory went back on the 12th of March to Rama's march from Ayodhya and to whom India's fight for freedom with the British Government is merely repetition of the fight of the Pandavas, the forces of righteousness, with the Kauravas, the forces of unrighteousness.
>
> Again there were no cinema films on the day when the Lord Buddha started out from his princely mansion leaving his dearest

ones behind, "roused into activity like a man who is told that his house is on fire". And yet to millions upon millions of men and women the story of Great Renunciation is as fresh as if it had taken place only yesterday.

And to come to modern times, there were no cinematographs on the 4th of July 1776 nor on the 14th of July 1789, and yet is there an American whom the memory of the 4th of July does not thrill with patriotic emotion, and is there a Frenchman whose heart does not swell with pride to think of the date on which the people free from the trammels of ages stormed the Bastille?

To come to recent times, the 13th of April lives in every Indian's memory without General Dyer having allowed any cinema company to take a photographic film on his bloody exploit.

Far different from any of the historical events we have noted, and fraught with consequences of much greater and more abiding import for the whole world is the struggle that was begun with the historic trek of the 12th of March. If it was a foolish adventure, none would be so foolish as to give a moment of thought to it; if it was a mere comedy, the censoring of the comic films cannot be characterized as better than an act of stupidity; if it was an event of great consequence both for India and the world, as it undoubtedly was, it was because of its inherent qualities, not because a few cinema companies regarded it as a profitable investment. Censorship or no censorship, ridicule or repression, the fight begun on the 12th of March is a fight to the finish, and the 12th of March will live in history as much more sacred and memorable than any other great date before, inasmuch as on that day began a march for freedom inaugurated by one who will more readily immolate himself and his band in the cause of truth and non-violence than directly or indirectly participate in the shedding of a single drop of blood.

Contrary to what Irwin had said in his telegram to Whitehall, considerable excitement was generated through the country by the march. Congress leaders in Calcutta appointed, with immediate effect, a Bengal Civil Disobedience Council. The council started a campaign to enrol volunteers for the Civil Disobedience Movement. In Bombay, the 'War Council' of the Provincial Congress Committee came out with a long list of volunteer satyagrahis. In Madras, the local District Congress Committee, the Andhra Congress Committee, the Triplicane Congress Sabha and the Young League got together so they could join the Civil Disobedience Movement.

Similar reactions also came from Lahore and from Peshawar, from Ahmedabad and Nagpur. In Allahabad, Jawaharlal Nehru hoisted a Swaraj flag on the AICC and City Congress offices, a task performed in Delhi by Devdas Gandhi who addressed a crowd of as many as ten thousand people. The march was beginning to have the effect Gandhi wanted. And it was just the start of the long walk to Dandi.

In the meantime, the marchers themselves were bravely soldiering on and, often, having to literally think on their feet. This was because the Dawn Brigade, zipping ahead of the marchers on any form of transport available, barely had time to make arrangements for the satyagrahis; it was natural that they couldn't foresee all eventualities.

For example, no one, not even in their most optimistic moments, anticipated the number of people who would line the streets, waiting for the marchers. Their number was in the thousands, so large that even *The Times of India* and other British newspapers, couldn't play down the crowds as they normally did. 'One hundred thousand people crowded the streets between the ashram and Ahmedabad city' was the common consensus. The crowd not only waited and cheered, but many joined in behind the satyagrahis, marching with the marchers in a show of solidarity. That often made the procession nearly a kilometre long.

Obviously, this had the effect of slowing down the march. Then someone had the idea of clearing a path for Gandhi by forming a human circle around him, volunteers linking hands to form a *cordon sanitaire*. The story goes that a plainclothes CID sleuth, sent by officialdom to keep tabs on the marchers from within, got his hands linked and became part of the protective circle! Whether his report that night included this nugget is not on record.

His report would definitely have mentioned that the marchers were fully expecting to be arrested in the first few minutes, because barely three kilometres from Sabarmati Ashram stood their replacements. As we have seen, there were nationalist students and they waited outside their alma mater the Gujarat Vidyapith. The two sets of marchers, original and replacements, passed each other silently in the morning.

At Ellis Bridge, the narrow structure which spanned the Sabarmati River, there was a real problem: the bridge was packed with spectators. So much so that there was no space for the marchers they had all come to see. Who would manage the crowds? No one had a solution till someone had his Eureka! moment by thinking out of the box: why not cross the river without using the bridge? Echoes of Jesus? It wasn't, because the river was almost dry, at its deepest only knee high. Dhotis were tucked in, chappals removed and held aloft, and the procession waded in, this time to cheers from all over, especially from the bridge.

From there, avoiding the centre of Ahmedabad, the marchers moved toward Chandola Lake. At the best of times, Ahmedabad is dry and dusty; the route to Chandola was particularly so, and the swirling winds, with the help of Western journalists' vehicles, caked everyone with dust. One report says that Gandhi covered his head with a sheet, whether to keep dust, or journalists, away no one can say.

Gandhi once again showed his contrarian side. Go home, he told the crowd which had been following the marchers. Their duty, according to him, was to continue with their own work. At least for the present. Their time for renunciation and sacrifice would come later, presumably when mass Civil Disobedience began.

Having thus got rid of their noisy bandwagon, Gandhi & Co walked on towards Aslali, which was to be their first halt. Along the route through villages informal reception committees waited to cheer the satyagrahis. At Aslali, which they reached at 10.20 a.m. after four hours of more or less continuous walking, there was a formal reception committee, complete with village dancers and accompanying musicians. Behind the committee, there was the inevitable crowd.

After the usual greeting of tilak and coconuts, the marchers rested under a shamiana that had been erected. Reports mention that dwellings and other buildings had been spruced up, something which was to become a regular feature of the 25-day march. There were no ornate decorations; everyone was aware of the Mahatma's commandments and no one dared to transgress them, but cleanliness he valued very highly.

Something else which was also to become a regular feature of the long march, was even more remarkable. As we have seen, Gandhi was leading a group of young men, most of them in their early twenties. He was sixty. He wasn't in the best of health. He was frail. He weighed a mere 45 kg. Yet he walked at the head of the column, leading at such a brisk pace that the youngsters found it difficult to keep up. Gandhi asked for no concessions: he carried his bags like everyone else, and when offers to carry them were made, Gandhi reacted angrily. In addition, he worked at his spinning wheel, wrote letters and gave speeches.

All this can be summarized in the story of the horse. An admirer in Ahmedabad had gifted it for Gandhi to ride on when he got too tired to walk. The gentleman knew Gandhi, but obviously didn't know him enough: Gandhi refused to ride it. So the horse brought up the rear of the column, an unnamed, eightieth marcher on the road to Dandi. Gandhi wouldn't even sling his bags over the animal's back to take the weight off his own shoulders.

At 4 o'clock, Gandhi spoke to the crowd of five thousand most of whom had waited patiently from the morning. This speech, with minor variations, was to be another regular feature of the march: the crowds expected Gandhi to speak, in fact, many had come expressly for that purpose. In any case, Gandhi, the Great Communicator, never lost a chance to spread the message of the Civil Disobedience Movement. His speech:

> Do not be content with merely wearing khadi and plying the spinning wheel, thinking that you have done all that you could do.
>
> Take the case of your own village: For a population of 1700, the salt required will be 850 maunds. For 200 bullocks, 300 maunds of salt will be required. That is, total of 1,150 maunds of salt will be required.
>
> The government levies a tax of Rs 1-4 on one pukka maund of salt. Hence, on 1,150 maunds, which is equal to 575 pukka maunds, you pay a tax of Rs 720.
>
> Can your village afford to pay this amount in taxes every year? In India, the average income of an individual is calculated at 7 pice or, in other words, hundreds of thousands of persons do not earn even a single pice and either die of starvation or live by begging. Even they cannot do without salt. What will be the plight of such persons if they can get no salt or get it at too high a price?
>
> Salt, which sells at 9 pies a maund in the Punjab, salt of which heaps and heaps are being made on the coast of Kathiawar and Gujarat, cannot be had by the poor at less than Re. 1-8-0 a maund. What curses the government may not be inviting upon itself from the poor for hiring men to throw this salt into mud!
>
> The poor destitute villagers do not have the strength to get this tax repealed. We want to develop this strength.
>
> A democratic State is one which has authority to abolish a tax which does not deserve to be paid. It is one in which the people can determine when a certain thing should or should not be paid.

We, however, do not possess such authority. Likewise, even our supposedly great representatives do not have it. In the Central Legislative Assembly, Pandit Malaviya said that the manner in which Sardar Vallabhbhai was arrested could not be called just; that it was unjust and high-handed. And this resolution was supported by Mr Jinnah. To this the government official replied that their magistrate had acted in a manner which befitted a loyal subject, if he had acted otherwise, he would have been regarded as traitor. If, however, that is the case, this bearded person (Abbas Saheb) and I should also be arrested, because I on my part openly make speeches about preparing salt.

We want to establish a government which will be unable to arrest a single individual against the wishes of the people, which cannot extract ghee worth even a quarter pice from us, cannot take away our carts, cannot extort money from us.

There are two ways of establishing such a government: that of the big stick or violence and that of non-violence or Civil Disobedience. We have chosen the second alternative, regarding it as our dharma. And it is because of this that we have set out to prepare salt after serving notice on the government to that effect.

I can understand there being a tax on such things as the hookah, bidis and liquor. And if I were an emperor, I would levy with your permission a tax of one pie on every bidi. And if bidis are found too expensive, those addicted to them may give them up. But should one levy a tax on salt?

Such taxes should now be replaced. We should make a resolve that we shall prepare salt, eat it, sell it to the people and, while doing so, court imprisonment, if necessary. If, out of Gujarat's population of ninety lakhs, we leave out women and children, and the remaining thirty lakhs get ready to violate the salt tax, the government does not have enough accommodation in jails to house so many people. Of course, the government can also beat up and shoot down those who violate the law. But the governments of today are unable to go to this extent. We, however, are determined to let the government kill us if it wishes.

> The salt tax must be repealed now. The fact that a sea of humanity had gathered and showered blessings upon us – for a distance of seven miles from the ashram to the Chandola Lake – a sight for the gods to see – that is a good omen. And, if we climb even one step, we shall readily be able to climb the other steps leading to the place of Independence.

The speech was followed by a collection of funds from the crowd as well as local committees. Then came the enlisting of volunteers to the cause, people who pledged to work for the movement. Last, and probably most difficult, there were the resignations: the local patil and peons, locals who had these coveted government jobs, voluntarily giving them up as a sign of protest against the rule of the British. This called for immense sacrifice. And willpower, because they were offered all kinds of inducements (like promotions and raises) to stay on in the job.

Dinner time – or more accurately, frugal meal time – was 5.30 p.m., followed by the spinning of charkhas and evening prayers. Lights went out at 9 p.m. for all except Gandhi, who was busy with interviews and writing. Wake up time, as usual, was 4 a.m. The marchers thus slept seven hours, Gandhi five.

The second halt was at Bareja, a village of 2500 people. From all accounts every man, woman and child turned up to see Gandhi and his satyagrahis. His speech to them can only be described as stern:

> This is our second halt after the march began. As at our first halt, here too I was given the required information about this village. I was pained to read it. It is strange that a place so near Ahmedabad has zeroes against the columns for consumption of khadi, the number of habitual khadi-wearers and spinning wheels at work. During my tours of North and South India, I used to follow a rule, namely, that the barber cutting my hair should be a khadi-wearer. But here you keep yourselves far away from such a thing as khadi. Khadi is the foundation of our

freedom struggle. All like khadi, but people are nowadays deterred by the fear that those wearing khadi will have to go to jail and die. Bareja has not a single khadi-wearer, which is indeed a painful fact. There is a khadi store here and you can certainly remove this blot. We do not disown our mother because she is fat or ugly to look at and adopt another woman, more beautiful, to fill her place. Foreign cloth will never bring us freedom. I request you to renounce luxuries and buy khadi from this heap before you.

The marchers had arrived in Bareja around 8 a.m.; they left at 6 p.m. They covered around five to six kilometres per hour, generally aiming to walk early in the morning and soon after sundown in order to avoid the March sun, which in Gujarat was already getting hot. Some of the walking was pleasant, because of surrounding vegetation, but that was an exception; most of the time, the terrain was bare and dusty, the ground hard on the feet. Blisters were a common feature. When they became too painful to bear, there was only one solution: take off the footwear and walk barefoot.

The majority of marchers may have been young, and they may have got used to the strict regimen of Sabarmati Ashram, but they were not necessarily used to physical exertion on this continuous scale. Besides which, their rest periods weren't exactly luxurious, their relaxation time being spent in pandals, sleep being on a different floor every night on the bedding each marcher carried on his back. The food, at best, was frugal (though not necessarily in quantity: everyone got their fill of homely cooking). Everyone was aware of Gandhi's extreme intolerance of 'extravagance': his central message being that Satyagraha was a sacrifice, not a picnic, and the marcher should eat no better than the inhabitants of the villages they passed through. Thus when one of the marchers was offered an ice-cream, it was reported to Gandhi, and an explanation had to be provided for such indulgence.

Nothing, though, seemed to deter the marchers. Some of them fell ill, a combination of sun, physical effort, iffy food and unhygienic water. But those who were ill hid it from Gandhi for as long as possible because they did not want to be seen as weak; more than that, they certainly did not want to be sent back to Sabarmati Ashram before the march was over. Reaching Dandi was such an obsession that when one of the marchers got high fever, he tried to keep the fact to himself. He couldn't for long because he was diagnosed to have smallpox. Even then he begged Gandhi to let him keep marching. Gandhi was happy with this sense of determination and found a solution: the patient would stay back for a couple of days of treatment, and then travel by truck or train and catch up with the marchers. There were many others who wanted to join in, but were stopped by Gandhi: this march wasn't about adding numbers. Even more remarkable was, of course, the willpower of Gandhi. The punishing schedule of the marchers was nothing compared to the brutal schedule Gandhi set himself. Or was forced to set himself, because everyone wanted a piece of him: he was the leader, the organizer, the trouble-shooter, the politician, the communicator, the dietician, the philosopher, all rolled into one underweight, rheumatic body.

So it went on for twenty-five days, covering a distance of 385 km and passing through forty villages and towns. These twenty-five days were battles within battles: there was the battle of the marchers' determination versus their bodies. There was the battle of propaganda: the nationalist press versus the British press. The much more widely circulated British press played down crowd numbers, underplayed the halts like Surat where hundreds of thousands of people arrived while gleefully reporting other halts where smaller crowds greeted the marchers. The nationalist press did the exact opposite.

There was also the battle between local administrations and the Civil Disobedience Movement: one of the objectives of the movement, as we have seen, was to generate a feeling which would make Indians working for the British government quit their jobs. There were village level patils and peons and magistrates and others. Many resigned because they felt a surge of nationalism, some resigned out of peer pressure, some were coerced due to pressure from nationalist friends, but all were discouraged and threatened and wooed in turn by the administration to stay.

The last, and most important was the battle of wits between Gandhi and Irwin. This started well before the march began with Gandhi's letter, and the Viceroy's options were many. Perhaps too many because a multiplicity of options often results in indecision. The first option was to prevent the march happening at all; to nip it in the bud at Sabarmati Ashram. A second was to arrest Gandhi the moment he preached sedition (which he did from Day One). A third was to arrest him on Day Two. A fourth, Day Three and so on.

In the event, Gandhi was allowed to be a free man till well after Dandi. This was, as we have seen, primarily because the British thought of the Salt March as a bit of a joke and expected the whole enterprise to collapse in ridicule. Once the march took on a life of its own and reached critical mass, mainly because Gandhi had picked in salt a symbol which cut across caste, creed and economic status, the moment had passed: arresting Gandhi would then have had the very effect he wanted: to galvanize the entire population.

So Irwin waited to arrest Gandhi. And Gandhi waited to be arrested by Irwin. When the arrest didn't happen as expected, a certain desperation began to creep into Gandhi's mind. How would he keep up the momentum over so many days? Could public enthusiasm be retained for so long? Wouldn't the sameness of the march, day after day, result in a jaded, and bored, public? Gandhi needed to be arrested, and when it didn't happen, his speeches became more and more strident in preaching sedition.

As we have seen, the march had fallen into its own rhythmic pattern of walking, rest and a Gandhi speech at each halt. Many of the speeches were variations of the speeches already reproduced here, but some of them were different and addressed issues which went beyond the immediate.

Gandhi described his speech at Bhatgam on 29 March as 'Turning the Searchlight inward'.

> Only this morning at the prayer time I was telling my companions that as we had entered the district in which we were to offer Civil Disobedience, we should insist on greater purification and intenser dedication. I warned them that as the district was more organized it contained many likelihoods of our being pampered. I warned them against succumbing to their pampering. We are not angels. We are very weak, easily tempted. There are many lapses to our debit. God is great. Even today some were discovered. One defaulter confessed his lapse himself whilst I was brooding over the lapses of the pilgrims. I discovered that my warning was given none too soon. The local workers had ordered milk from Surat to be brought in a motor lorry and they had incurred other expenses which I could not justify. I therefore spoke strongly about them. But that did not allay my grief. On the contrary it increased with the contemplation of the wrongs done.

THE RIGHT TO CRITICIZE

> In the light of these discoveries, what right had I to write to the Viceroy the letter in which I have severely criticized his salary which is more than five thousand times our average income? How could I (object)...? Certainly not if I was myself taking from the people an unconscionable toll. I could resist it only if my living bore some correspondence with the average income of the people. We are marching in the name of God. We profess to act on behalf of the hungry, the naked and the unemployed. I have no right to criticize the Viceregal salary, if we are costing the country say fifty times seven pice, the average daily income of our

people. I have asked the workers to furnish me with an account of the expenses. And the way things are going, I should not be surprised if each of us is costing something near fifty times seven pice. What else can be the result if we would take all the dainties you may place before us under the excuse that we would hurt your feelings, if we did not take them. You give us guavas and grapes and we eat them because they are a free gift from a princely farmer. And then imagine me with an easy conscience writing to the Viceroy a letter on costly glazed paper with a fountain pen, a free gift from some accommodating friend!!! Will this behove you and me – can a letter so written produce the slightest effect?

... Therefore in your hospitality towards servants like us, I would have you to be miserly rather than lavish. I shall not complain of unavoidable absence of things. In order to procure goat's milk for me you may not deprive poor women of milk for their children. It would be like poison if you did. Nor any milk and vegetables be brought from Surat. We can do without them if necessary. Do not resort to motor cars on the slightest pretext. The rule is, do not ride, if you can walk. This is not a battle to be conducted with money. It will be impossible to sustain a mass Movement with money. Anyway it is beyond me to conduct the campaign with a lavish display of money.

... We may not consider anybody as low. I observed that you had provided for the night journey a heavy kitson burner mounted on a stool which a poor labourer carried on his head. This was a humiliating sight: This man was being goaded to walk fast. I could not bear the sight. I therefore put on speed and outraced the whole company. But it was no use. The man was made to run after me. The humiliation was complete. If the weight had to be carried, I should have loved to see some one among ourselves carrying it. We would then soon dispense both with the stool and the burner. No labourer would carry such a load on his head. We rightly object to *beggar* (forced labour). But what was this if it was not *beggar*? Remember that in Swaraj we would expect one drawn from the so-called lower class to preside over India's destiny. If then we do not quickly mend our ways, there is no Swaraj such as you and I have put before the people.

From my outpouring you may not infer that I shall weaken in my resolve to carry on the struggle. It will continue no matter how co-workers or others act. For me there is no turning back whether I am alone or joined by thousands. I would rather die a dog's death and have my bones licked by dogs than that I should return to the ashram a broken man.

Gandhi wasn't content just saying this piece. He announced that as a penance, he would, for the rest of the march, eat a restricted diet. He would eat no fresh fruit or fruit juices. He would only have dates, currants and lime juice.

A few days earlier, Gandhi had also spoken on the contentious issue of Hindu-Muslim relations. His speech began with a question:

A Muslim youth has sent me questions on the Hindu-Muslim problems. One of them is, "Do you expect to win Swaraj through your own single effort or assisted merely by the Hindus?" I have never dreamt that I could win Swaraj merely through my effort or assisted only by the Hindus. I stand in need of the assistance of Musalmans, Parsis, Christians, Sikhs, Jews and all other Indians. I need the assistance even of Englishmen. But I know too that all this combined assistance is worthless if I have not one other assistance that is from God.

But to realize His help and guidance in this struggle, I need your blessings, the blessings of all communities. The blessings of thousands of men and women belonging to all communities that have attended this march are to me a visible sign of the hand of God in this struggle ...

I am thankful to be able to say that I have had during the march abundant proof of the blessings of these communities. I have seen real friendliness in the eyes and in the speech of the

Musalmans who along with the rest have lined our route or attended the meetings. They have even given material aid.

As planned, the marchers had reached Dandi on 5 April the day before D-day, 6 April 1930. Gandhi spent the day giving interviews and in meetings with Congress leaders because he was certain that he would finally be arrested immediately after he broke the salt law. In the evening a large crowd – large for distant, uninhabited, Dandi – had gathered to hear Gandhi speak, his probable last speech as a free man for some time to come. He said:

> When I left Sabarmati with my companions for this seaside hamlet of Dandi, I was not certain in my mind that we would be allowed to reach this place. Even while I was at Sabarmati there was a rumour that I might be arrested. I had thought that the government might perhaps let my party come as far as Dandi, but not me certainly. If someone says that this betrays imperfect faith on my part, I shall not deny the charge. That I have reached here is in no small measure due to the power of peace and non-violence; that power is universally felt. The government may, if it wishes, congratulate itself on acting as it has done, for it could have arrested every one of us. In saying that it did not have the courage to arrest this army of peace, we praise it. It felt ashamed to arrest such an army. He is a civilized man who feels ashamed to do anything which his neighbours would disapprove. The government deserves to be congratulated on not arresting us, even if it desisted only from fear of world opinion.
>
> Tomorrow we shall break the salt tax law. Whether the government will tolerate that is a different question. It may not tolerate it, but it deserves congratulations on the patience and forbearance it has displayed in regard to this party.
>
> If the Civil Disobedience Movement becomes widespread in the country and the government tolerates it, the salt law may be taken as abolished. I have no doubt in my mind that the salt tax

stood abolished the very moment that the decision to break the salt laws was reached and a few men took the pledge to carry on the Movement even at the risk of their lives till Swaraj was won.

If the government tolerates the impending Civil Disobedience you may take it for certain that the government, too, has resolved to abolish this tax sooner or later. If they arrest me or my companions tomorrow, I shall not be surprised, I shall certainly not be pained. I would be absurd to be pained if we get something that we have invited on ourselves.

What if I and all the eminent leaders in Gujarat and in the rest of the country are arrested? This Movement is based on the faith that when a whole nation is roused and on the march no leader is necessary. Of the hundreds of thousands that blessed us during our march and listened to my speeches there will be many who are sure to take up this battle. That alone will be mass Civil Disobedience.

We are now resolved to make salt freely available in every home, as our ancestors used to, and sell it from place to place, and we will continue doing so wherever possible till the government yields, so much so that the salt in government stocks will become superfluous. If the awakening of the people in the country is true and real, the salt law is as good as abolished.

But the goal we wish to reach is yet very far. For the present Dandi is our destination but our real destination is no other than the temple of the goddess of Swaraj. Our minds will not be at peace till we have her darshans, nor will we allow the government any peace.

The night couldn't have been easy for the marchers. But Gandhi, as usual went to sleep at the appointed time as if the next morning was just another day at the office.

At 4 a.m. everyone was up, and sometime before six, morning prayers behind them, they were all at the beach. Gandhi entered the sea first for a symbolic dip in the water. Then at 6.30 a.m., he bent down and picked up a fistful of salt.

'With this salt I am shaking the foundations of the empire,' are his legendary words. Sarojini Naidu who was standing closest to Gandhi, said 'Hail Deliverer!' after which everyone shouted, 'Mahatma Gandhi ki jai' and *Vande Mataram*.

The marchers now all rushed in for their moment of triumph, each one scooping up in lotas what was a mixture of wet sand and salt, then depositing it in large pans for boiling. This went on for most of the morning, as they made one trip after another to gather more raw material. The crowd which had been waiting from before dawn and had been cheering wildly, also joined in in the salt-making. At the end of the day, a thousand pounds of salt had been obtained. Gandhi's 'two tolas' were of course, the most precious part of the day's collection and later travelled across the whole country to be auctioned and re-auctioned to raise money for the Swaraj fund.

That first lump of salt picked up by Mahatma Gandhi was the signal for salt to be lifted from every natural source in the country. Thus Gandhi, the first law-breaker, was joined by hundreds of thousands of law-breakers to start a nation-wide Civil Disobedience Movement. Would it lead to Complete Independence? Only time would tell. But the first step towards freedom had certainly been taken at the salt pans of Dandi.

EPILOGUE

6 April 1930 may have marked the end of the Dandi Salt March, but it was only the beginning of the Civil Disobedience campaign. The campaign took off, without a shadow of a doubt, because Gandhi's instinct had been right, and in choosing the breaking of the salt laws and the long march to Dandi, he had captured the public imagination once again.

What followed Gandhi's first scoop of a fistful of salt was magical: all over India there was enthusiastic participation in the breaking of the salt laws as well as other laws which were deemed unjust and extortionist. The participants weren't just committed satyagrahis; ordinary people, many for the first time, felt a surge of nationalism which the authorities found difficult to contain: one estimation is of five million people gathering together at five thousand meetings throughout the country.

The raids on government salt works which followed, at Dharasana and in Wadala, involved many ordinary volunteers (and many spectators, some of whom were moved to participate in the raids). They would walk quietly in the area around the salt pans, each carrying an empty sack of moderate size. At some point, each would make a dash for the pans, scoop as much salt into the sack and make a run for it. Indian police constables with their European officers, who had been watching till then, then sprang into action, seizing the sacks and arresting the volunteers, who did not put up any resistance. At this, yet another wave of volunteers would come up and replace the earlier batch and keep the action going. The scene was to be repeated at Wadala with one difference: the police did not wait for the volunteers to fill up their sacks. They rained lathi blows on them in the act of collecting. Many of the volunteers took the blows on their bodies but continued collecting salt until they were hauled up and put into police vans, in their fortitude and courage setting an example for others to follow. And inspire many others to join the Movement for Swaraj.

We now know that Independence didn't come to India until a full seventeen years later, so, in a way, the Dandi Salt March didn't succeed in its stated objective.

But the Dandi March and the Civil Disobedience Movement which followed, achieved much else. To start with, it gave a sense of empowerment to the rural poor, the sense that it was possible even for them to confront the might of the British Empire, and get away with it. That message, put across directly as well as subliminally, got through to more and more people, almost all of them rural, almost all of them poor. It was noteworthy that in village after village, the entire population would turn out to hear Gandhi speak. The message of empowerment wasn't accidental: Gandhi wrote about it extensively. Civil Disobedience, he said, was not just 'designed to establish independence but to arm the people with the power to do so.' Elsewhere he wrote 'Once his (satyagrahi's) mind is rid of fear, he will never agree to be another's slave.' Swaraj, he seemed to say, was a process of freeing oneself. The rest would follow. It wasn't only the poor he was trying to reach: in the call for resignations of government office bearers, he was asking people to put the cause above themselves. And his rich supporters risked government action when they supported Gandhi. In this

Gandhi was going beyond the idea of Swaraj; he was sowing the seeds of a social revolution.

The Dandi Salt March also marked a period in which the concept of Satyagraha was put into practice on a much larger, and more widespread way than ever before, so that it wasn't just Gandhi's close acolytes who practised it, but a wider circle of committed nationalists who became aware that the means to an end were as important as the end. 'Satyagraha,' Gandhi wrote just before the march began, 'arms people with power not to seize power but to convert the usurper to their own view till at last the usurper retires or sheds the vices of a usurper and becomes a mere instrument of service of those whom he has wronged. The mission of satyagrahis ends when they have shown the way to the nation to become conscious of the power lying latent in it.'

It is not an easy concept to understand, and much less easy to put into practice. Yet it was possible, during and after the Dandi Salt March to see Satyagraha change the way the British, and the world began to see India and Indians. There was growing respect for nationalist sentiments, and Lord Irwin, particularly, began to see Gandhi as an equal, which, given the attitudes and power structure of those days, was an astonishing transformation. (Winston Churchill, then leader of Britain's Opposition, was plainly disgusted. 'This naked fakir,' he had said infamously, meeting on equal terms with the Viceroy of His Majesty's government!)

The aftermath of Dandi also began to change world opinion. Although the British press, both in India and in the home country, was completely biased, the Dandi Salt March brought in a more objective American media. Its attitude, sneering at first, began to change when reporters saw Satyagraha and non-violence in action against an imperial power which did not hesitate to use physical force.

William Shirer was one of the leading American correspondents whose views became increasingly sympathetic to the national cause, influenced public opinion back home. Another was Webb Miller whose reports from the 'front' were graphic yet objective, a potent combination. The following is a report of the Dharasana salt raid, soon after Gandhi's arrest and detention in jail:

> Slowly and in silence the throng commenced the half-mile march to the salt deposits. A few carried ropes for lassoing the barbed-wire stockade around the salt pans. About a score who were assigned to act as stretcher-bearers wore crude, hand-painted red crosses pinned to their breasts; their stretchers consisted of blankets. Manilal Gandhi, second son of Gandhi, walked among the foremost of the marchers. As the throng drew near the salt pans they commenced chanting the revolutionary slogan, *Inquilab zindabad*, intoning the two words over and over.
>
> The salt deposits were surrounded by ditches filled with water and guarded by four hundred native Surat police in khaki shorts and brown turbans. Half a dozen British officials commanded them. The police carried lathis – five-foot clubs tipped with steel. Inside the stockade twenty-five native riflemen were drawn up.
>
> In complete silence the Gandhi men drew up and halted a hundred yards from the stockade. A picked column advanced from the crowd, waded the ditches, and approached the barbed-wire stockade, which the Surat police surrounded, holding their clubs at the ready.
>
> Suddenly, at a word of command, scores of native police rushed upon the advancing marchers and rained blows on their heads with their steel-shod lathis. Not one of the marchers even raised an arm to fend off the blows. They went down like ten-pins. From where I stood I heard the sickening whacks of the clubs on unprotected skulls. The waiting crowd of watchers groaned and sucked in their breath in sympathetic pain at every blow.
>
> Those struck down fell sprawling, unconscious or writhing in pain with fractured skulls or broken shoulders. In two or three

minutes the ground was quilted with bodies. Great patches of blood widened on their white clothes. The survivors without breaking ranks silently and doggedly marched on until struck down. When every one of the first column had been knocked down stretcher-bearers rushed up unmolested by the police and carried off the injured to a thatched hut which had been arranged as a temporary hospital.

Then another column formed while the leaders pleaded with them to retain their self-control. They marched slowly towards the police. Although every one knew that within a few minutes he would be beaten down, perhaps killed, I could detect no signs of wavering or fear. They marched steadily with heads up, without the encouragement of music or cheering or any possibility that they might escape serious injury or death. The police rushed out and methodically and mechanically beat down the second column. There was no fight, no struggle; the marchers simply walked forward until struck down. There were no outcries, only groans after they fell. There were not enough stretcher-bearers to carry off the wounded; I saw eighteen injured being carried off simultaneously, while forty-two still lay bleeding on the ground awaiting stretcher-bearers. The blankets used as stretchers were sodden with blood.

The Gandhi men altered their tactics, marched up in groups of twenty-five and sat on the ground near the salt pans, making no effort to draw nearer. Led by a coffee-coloured Parsi sergeant of police named Antia, a hulking, ugly-looking fellow, detachments of police approached one seated group and called upon them to disperse. The Gandhi followers ignored them and refused even to glance up at the lathis brandished threateningly above their heads. Upon a word from Antia the beating recommenced. Bodies toppled over in threes and fours, bleeding from great gashes on their scalps. Group after group walked forward, sat down, and submitted to being beaten into insensibility without raising an arm to fend off the blows.

Finally the police became enraged by the non-resistance.

> They commenced savagely kicking the seated men in the abdomen and testicles. The injured men writhed and squealed in agony, which seemed to inflame the fury of the police. The police then began dragging the sitting men by the arms or feet, sometimes for a hundred yards, and throwing them into ditches. Hour after hour stretcher-bearers carried back a stream of inert, bleeding bodies.

Those inert, bleeding bodies, some of them already dead, others to die soon after, stirred the conscience of American readers and American leaders. In later years, Roosevelt often brought up the 'Indian question' with Churchill, an intrusion Churchill resented, but couldn't ignore. Those inert, bleeding bodies also began to creep into British consciousness and thus into British consciences, in spite of the initial British press hostility. The process of Gandhi becoming a hero to the British working class must have also begun at this very time.

There were other messages from the Dandi Salt March. The most important was of Inclusion: Gandhi tried to bring in Muslims, Untouchables and women into the mainstream of action. There was, as we have seen, only limited success with Muslims, though, even here, Gandhi used his instinct for symbolism. 'In Dandi,' Gandhi announced, 'a Muslim has invited me and I will be putting up at his bungalow. Satyagraha will commence from the Muslim friend's home.'

With Untouchables, it would be a long drawn-out battle, one which continues even today, but the battle was at least begun. The change in today's attitudes has a lot to do with the forceful example Gandhi set during the course of the march.

Thomas Webber describes this encounter which occurred on the tenth day of the march at a village called Gajera:

> Under a large banyan tree a dais had been erected for Gandhi and at 3.45 he took his place for the commencement of his speech. Four to five thousand people sat patiently waiting for the

Mahatma to tell them about the iniquities of the salt tax, the evils of British rule and the meaning of Swaraj. Instead, Gandhi just sat on the platform and waited. The usual hymn from Pandit Khare was not forthcoming. Silence increased the growing tension. For the first time during the march the "untouchables" of a village had been prohibited from sitting with the rest of the audience. Gandhi told his volunteers to sit among the Harijans and finally he announced: "This meeting has not yet started ... Either you invite the 'untouchables' and my volunteers to sit freely among you or I'll have to address you from the hill where they are sitting." He then awaited an answer.

Gandhi had already known, before his speech, that the Harijans would not be in the audience. The village elders did not want them integrated and the Harijans themselves were afraid of the possible consequences if they took the insolent action of making the first move. The displeased Gandhi decided to set social interactions at Gajera on the right track.

One version of the outcome of this confrontation is a success story in the true vein of Gandhian conflict resolution through conversation. After a further period of silence an elder finally rose and announced that desegregation was fine by him. Another elder, then another and another rose in turn to affirm the position of the first, until it appeared that they all wanted the "untouchables" in their midst.

All this time the women had said nothing. Gandhi turned to them, whereupon one woman said that they had no objections. She added that "it is the menfolk who want to perpetuate segregation". Gandhi then invited the "untouchables" to join the group. This, once everyone had settled, led, in the words of one of the marchers, to "an emotional upsurge of religious rebirth". Gandhi "exhorted the high caste Hindus to promise themselves that this act of self-purification would remain with them permanently, that they would neither abuse nor neglect their underprivileged fellow citizens. The promise was made."

In the case of women, there is no doubt that the Dandi Salt March was the beginning of a revolution. *Young India* has a number of references to the role of women. 'The most gratifying feature of the receptions,' goes one report, 'was the part played by the women of the villages who came out of their purdah in their hundreds to receive the satyagrahis.' Or, 'Two lady workers of our province Shrimatee Rama Devi and Malti Devi who have wholeheartedly joined the present struggle, went to the first Satyagraha centre and in company with several women of the villages manufactured salt at two different places. Hundreds of women accompanied them in a procession, blowing conches.'

Although he had excluded women from the march, Gandhi encouraged them to take part in the Civil Disobedience Movement, particularly in picketing liquor outlets and shops selling foreign cloth. Women, of course, played an increasingly important role in the running of Sabarmati Ashram when most of its men went on the march. They sought to increase that role, and spread awareness of other roles women could play by starting classes at the ashram for future women satyagrahis.

But even Gandhi wasn't prepared for the speed with which women joined the movement in an age where purdah prevailed, and society was deeply conservative. In village after village, more and more women joined the crowds which greeted the marchers, with their proportion in the crowd going up from ten per cent to as much as a quarter. After the march terminated at Dandi, women took on work which Gandhi expected (like selling contraband salt under the name of 'Gandhi Salt' and 'Swaraj Salt'). And work which Gandhi hadn't expected (like goading policemen to take action against them, leading finally to a decision by the British to stop being 'chivalrous' to women).

The Dandi Salt March also brought non-violence into the battlefield, so to speak. As Gandhi said, 'We are constantly being astonished these days, at the amazing discoveries in the field of violence. But I maintain that far more undreamt of and seemingly impossible discoveries will be made in the field of non-violence.'

Many of those, of course, are still to be made. But some were made, and as Martin Luther King and Nelson Mandela would testify, the seeds were sown somewhere during the Salt March to Dandi.

Post-script: Months after the Dandi Salt March when Gandhi met Lord Irwin for negotiations, he took out a small paper bag from the folds of his chadar and dropped its contents into his cup. 'I will put a little of this salt into my tea,' Gandhi said with a mischievous smile, 'to remind us of the famous Boston Tea Party.' Lord Irwin laughed with him.

Part 2
SALT'S MARCH THROUGH HISTORY

THE STUFF OF LIFE

Salt. Another four-letter word for Life. We take it for granted, dismiss it as common salt, but without it, we couldn't exist. It's in our blood, in our sweat, in our tears. It's in semen and in urine. It's in our biological tissues and is a necessary component for the functioning of cells.

Salt is sodium chloride, NaCl. Sodium is needed to move nutrients in the body, to transport oxygen and to transmit nerve impulses (through signal transduction by sodium ions). It is essential for the movement of muscle (and that includes the heart). It maintains the electrolyte balance between the fluid inside and outside of cells; without this balance, osmotic pressure would cause our cells to either explode or dehydrate and shrivel. The body doesn't produce it, so we have to look for it. Its central role in life led many to call it the Fifth Element. Earth. Air. Fire. Water. And Salt. Which is why from the beginning of civilization, salt has been one of the most sought-after commodities on earth.

An adult human has 9 gm of salt per litre of blood. He has a total of 250 gm of salt inside his body. That's enough to fill four salt shakers. An adult loses some of it every day through sweat and urine and if it's not replaced, the repercussions are severe. Depending on the level and length of deprivation, the reactions can range from headache to light-headedness to nausea to even death in extreme cases. Strangely enough the body doesn't crave salt, even in the worst scenario, whereas the absence of water makes us feel parched, the absence of air suffocates us, if there's no light or heat, we feel the cold. Perhaps that's why we like a salty taste, and feel salt's absence in food immediately, whatever our culture and whatever our cuisine.

In early civilization, at the time when man was a hunter, he didn't need salt because he ate a lot of red meat, and all the sodium he needed came from the meat of carnivorous animals which were part of his diet.

Animals, incidentally, also need salt for the same reasons we do, but carnivorous animals get it from the animals they eat. If those animals were herbivorous, they didn't get nearly enough from their diet of plants and grass, so in the food chain, they were the ones who had to go looking for it. They found what were, literally, salt licks, salt deposits in rocks which they licked to get the required intake. Sometimes, this cycle got shorter, as when animals went around settlements and got their 'lick' of salt from human urine on the outskirts of the settlements.

As man became agrarian, his diet became more vegetarian, and he needed an intake of salt directly. To find salt sources, he often followed herbivorous animal trails till he came across salt deposits. Later, when he began to domesticate animals, he found that these animals too needed salt, and in greater quantities than he himself did. A horse, for example, needs five times as much as man, while a cow needs ten times the human level.

In short, salt literally gives life. In Hinduism, in fact, that's how the world began, with the churning of the primordial sea. Our seas teem with life, their fecundity seems limitless and their salty waters sustain our life.

Salt gave life in other ways. For example, by being a preservative. Nowadays refrigerators and cold storages have become so ubiquitous that it is difficult to imagine a time when they weren't there. As it happens, they came into use only recently. Before them, salting food heavily was the only way to keep it from decaying. Cod, tuna, sardines, herrings from the sea, meats of many kinds from land, were all salted in well-defined ways so they would keep for days. Without that, man wouldn't have gone far; sailors at sea for days on end, were particularly grateful for salt's preservative qualities. Winter months in Europe and other cold countries were made bearable because salt-preserved hams and olives and fish saw them through.

The combination of taste and preservation made salt a precious commodity. Since it wasn't as readily available everywhere as it is now, in some areas it was even used as currency. The Latin phrase *salarium argentums* (salt money) referred to the part payment in salt of a Roman soldier's wages. From that phrase has come the English word 'salary', which we now certainly don't use for salt. In Tibet, pieces of salt shaped in a mould served as change: eighty such pieces were equal in value to a saggio of fine gold (the Roman *solidus*). Marco Polo in his travels to China reported that salt was used as money in Yun Nan and other provinces of south-western China. In ancient Greece, slaves were traded for salt, and an unruly or incompetent slave was termed 'not worth his salt'. In Abyssinia, on the other hand, it wasn't currency: a guest was welcomed home by being offered a piece of rock salt which was expected to be licked.

The importance of salt is reflected in folk stories like the one which follows. This is a Roman one, but you find similar ones from almost all countries, with little or no variation.

THE VALUE OF SALT

A Roman Folk Tale

There was once a king who had three daughters. He was very keen to know which of them loved him most; he tried several tests but was never satisfied with the results.

One day he thought he would settle the matter once and for all, by asking each one of them the same question, then evaluating their answers. First he called the oldest by herself, and asked her how much she loved him.

'As much as the bread we eat,' was her reply; and he said to himself, 'She must love me the most of all; for bread is the first necessity of our existence, without which we cannot live. She means, therefore, that she loves me so much she could not live without me.'

Then he called the second daughter by herself, and said to her, 'How much do you love me?'

She answered, 'As much as wine.'

That is a good answer too, said the king to himself. It is true she does not seem to love me quite as much as the eldest; but still, scarcely can one live without wine, so that there is not much difference.

Then he called the youngest by herself, and said to her, 'And you, how much do you love me?'

She answered, 'As much as salt.'

The king was angry. 'What a contemptible comparison! She only loves me as much as the cheapest and commonest thing that comes to the table. This is as much as to say, she doesn't love me at all. I will never see her again.'

So he ordered that a wing of the palace should be shut up from the rest, where she should be served with everything belonging to her royal status, but where she should live by herself apart, and never come near him.

Here she lived, then, all alone. But though her father thought she did not care for him, she pined so much at being kept away, that at last she could bear it no longer.

The room that had been given her looked upon an inner courtyard. Here she sometimes saw the cook come out and wash vegetables at the fountain.

'Cook, cook!' she called one day, as she saw him pass under the window.

He looked up with a good-natured face, which gave her encouragement.

'Don't you think, cook, I must be very lonely and miserable up here all alone?'

'Yes, Signorina,' he replied; 'I often think I should like to help you to get out; but I dare not think of it, the king would be so angry.'

'No, I don't want you to do anything to disobey the king,' answered the princess; 'but would you really do me a favour, which would make me very grateful indeed?'

'Oh, yes, Signorina, anything which I can do without disobeying the king,' replied the faithful servant.

'Will you just oblige me so far as to cook papa's dinner today without any salt in anything? Not the least grain in anything at all. Let it be as good a dinner as you like, but no salt in anything. Will you do that?'

'Yes, I will,' said the cook.

That day at dinner the king had no salt in the soup, no salt in the boiled meat, no salt in the roast, no salt in the fried fish.

'What is the meaning of this?' thundered the king, as he pushed dish after dish away from him. 'There is not a single thing I can eat today. I don't know what they have done to everything, but there is not a single thing that has got the least taste. Let the cook be called.'

So the cook was brought before him.

'What have you done to the victuals today?' said the king sternly. 'You have sent up a lot of dishes, and no one alive can tell one from another. They are all of them exactly alike, and there is not one of them that can be eaten. Speak!'

The cook answered, 'Hearing your Majesty say that salt was the commonest thing that comes to the table, and altogether worthless and contemptible, I considered whether it was a thing that at all deserved to be served up to the table of the king; and, judging that it was not worthy, I abolished it from the king's kitchen, and dressed all the meats without it. Barring this, the dishes are the same that are sent every day to the royal table.'

Understanding now dawned on the king. He now realized how great was the love of his youngest child for him; so he had her apartment opened, and called her to him and embraced her with tears in his eyes.

The importance of salt led to the logical next step: governments began to earn revenue from it, either for taking over sources of salt, or by taxing its production and sale. The ancient Chinese taxed it and so did the Romans. The Chinese emperor Hsia Yu (2200 B.C.) was said to be the first to tax salt; in fact, it is likely that this was the very first tax imposed in history. The French slapped on a Salt Tax called *la gabelle* which was the last straw for an already seething population, and led to the French Revolution.

Since it was such a precious commodity, salt acquired a social status: the aristocracy started putting it on the dining table in salt cellars, which grew more and more elaborate and were made of silver and even more expensive materials. Not just that: the cellars were placed in the middle of the table, at the head of which sat the host. A guest's proximity to the host was relative to the cellar – whether he/she sat 'above the salt' or 'below the salt' – defined his or her social position.

Salt also acquired a metaphorical importance. The Jews' Covenant with God was of salt. The Catholic Church uses Holy Salt, *Sal Sapientia*, in many of its services. The fact that even if salt is dissolved in water, it can be evaporated back to its original crystalline state suggests longevity and permanence. Salt, therefore, became a symbol of friendship and of loyalty. Perhaps from this comes the widely used Hindi phrase which translates as 'I have eaten his salt' (so I will be loyal to him) and its opposite, 'Namak haram'.

In its wider usage, salt isn't just sodium chloride: in chemistry, salt is any chemical compound formed by the reaction of an acid with a base with the hydrogen of the acid replaced by metal or other cation. Thus magnesium chloride too is a salt as is potassium chloride. What is called saltpetre (or saltpeter) can be either potassium nitrate or sodium nitrate and was used by the Chinese as gunpowder.

Some 120 countries in the world produce salt. Of these, the United States of America is the largest, producing forty-five million tonnes per year. Its production is followed, in order, by China, India, Germany and Canada. The total salt production in the world is nearly 230 million tonnes. Being a bulky, low-value commodity, it is generally consumed near its production base and less than twenty per cent of world production is traded internationally. Surprisingly, only a small part of this total (about twenty per cent) has to do with the salt we eat. About ten per cent is used by the US to melt ice and snow on roads in winter (salt lowers the melting point). It is also used in water-softening systems. The chemical industry, in fact, is the biggest user of salt requiring half the salt produced.

The main supply for the United States comes from the Great Salt Lake, which is the fourth largest lake in the world and the Detroit Salt Mine which is 1200 feet below the city and covers an underground expanse of 1400 acres, connected by as many as fifty miles of road.

India has an average annual production of fifteen million tonnes. At Independence, as the first part of the book shows, there was artificially created scarcity and salt had to be imported. Soon after 1947, the Indian government fulfilled the pledge given by Mahatma Gandhi when he began the Salt March to Dandi, and abolished the tax on salt.

By 1953, in the short span of six years, self-sufficiency in salt production was achieved. From there, it was rapid progress to a position where India exported 32,500 tonnes of salt to the United States. Currently, a total of 1.5 million tonnes is exported to various countries every year.

Salt is produced in India in as many as 10,107 salt works, most of them in the small sector. The largest producer by far, is

Tata Salt, part of Tata Chemicals whose inorganic chemicals complex is a fully integrated production facility at Mithapur (the City of Salt), on the west coast of Gujarat. The Mithapur Salt Works are spread over 60 sq km and generate two million tonnes of solar salt, which is the starting raw material for the twenty-seven basic chemicals that Tata Chemicals manufactures. The four principal end uses of salt are the manufacture of chlorine and caustic soda, soda ash and edible salt. Chlorine, in turn, has many uses, from the treatment of water for drinking to serving as a base ingredient for PVC plastic.

The International Salt Commission actually counted the various uses of salt and came up with the astonishing figure of 14,000. Some of these uses are:

- To freeze ice-cream
- Removing rust
- Sealing cracks
- Maintaining colour of vegetables while boiling
- Removing stains on cloth
- Killing poison ivy
- For sore throats
- Melting ice from roads in winter
- Fertilizing farming fields
- Killing farm fungus
- Softening water
- Dying textiles
- PVC manufacture

One use not mentioned because it is not in current use, was the Egyptian art of mummifying the royal dead.

Salt's ability to seal cracks has come to an unexpected use: disused salt mines are now thought to be the best place to store nuclear waste which can remain toxic for a very long time, say 240,000 years. New Mexico salt mines have already been used for this purpose, and presumably more will be used as greater amounts of nuclear waste are generated and need to be buried.

Salt domes have also been used to stockpile the Strategic Oil Reserve, an emergency reserve of petroleum stored by the United States for use in times of war or when the US is under attack. As many as 700 million barrels of oil have been stored in five hundred salt domes in Texas and Louisiana.

The most unusual, and certainly the most expensive goods stored in a salt mine according to Kurlansky, was in Germany, when that country was defeated and its troops were on the run. In 1945, soon after World War II ended, American troops came across a salt mine where valuables were 'salted' away: 100 tonnes of gold, rows and rows of gold coins and international currency worth two million dollars! That isn't all: perfectly preserved by the low, steady humidity and the even, cool temperature of the salt mine were paintings of the Masters. Like Rembrandt and Raphael. The total booty stashed 1200 feet under the surface and found accidentally, was a cool three billion dollars.

Most of us are familiar with refined salt, which is generally what is served on our table. But there are a number of unrefined salts, with their own individual characteristics. For example, Normandy and Brittany Salts are grey (*moist sel gris*) and off-white with lacy flakes and a slight sweetness (*fleur des sel*, often called 'the champagne of sea salt'). Hawaiian salt is pinkish from clay containing iron oxides. Korean bamboo salt is sea salt poured into bamboo cylinders which are plugged into clay and roasted over burning pine resin. Obviously, its taste is like no other. There's also kosher salt, rock salt mostly used for pickling. Cordon Bleu chefs often use these salts in signature dishes to get a distinctive flavour. As it happens, unrefined salts also have more minerals than refined salt, though that advantage is sometimes cancelled by the fact that the former have a colour which may be due to sand or dirt.

A disciple of Mahatma Gandhi once pointed out (though presumably not to him), that *himsa* (violence) was the norm in the animal world where we too have come from. He was right, of course: Civilization's many facets may have subdued our most violent instincts, but violence remains central to what we eat. *Jeevo Jeevasya Jeevanam:* Life feeds on life. Even a vegetarian eats things which have life. The only non-organic thing we eat is a rock we call salt. That, in itself, makes salt truly unique.

SALT THROUGH HISTORY

The history of salt begins with the ocean. Geologists believe that inland salt deposits (rock salt or halite) originated as the residue of sea water that became enclosed by land millions of years ago, then evaporated. This resulted in beds of salt. Some of them remained on the surface; others got buried over the centuries under layers of rock from other types of sedimentation. For example, 400 million years ago, the sea covering the area now known as Michigan evaporated, leaving behind huge salt deposits which got buried by glacial activity. This salt has spread underground, over 270,000 sq km below Michigan, Ontario, Ohio, Pennsylvania, New York and West Virginia. There is enough salt there, it is estimated, to last seventy million years. That should see us through for a while.

The cycle is completed by rivers that flow into the ocean carrying minerals from land and rock which include positive ions of sodium, calcium and potassium. Volcanoes on the ocean floor give out hydrogen and negative chloride ions. These get paired through chemical and biological processes to give salts.

Ocean water contains 3.5 per cent salts of which a large portion is NaCl, our common salt or sodium chloride. The balance is made up of magnesium, sulphur and calcium. Sea water, in fact, contains all the elements that make up the minerals in the earth's crust.

(There is an interesting parenthesis here. It has been established that the composition of sea water is identical to the composition of body fluids in man and animal, i.e., the proportion of elements is much the same. This seems to support the theory that life began in the ocean.)

Salt beds on earth range from a few feet in thickness to several hundred feet. Subterranean salt beds sometimes become domes because pressure from surrounding rock strata forces them upwards.

The earliest known writing on salt goes back some 4700 years ago. In around 2700 B.C. in China, Peng-Tzao-Kan-Mu wrote a treatise on pharmacology, a major portion of which was devoted to a discussion of more than forty kinds of salt. The book also included two methods of extracting salt and getting the product into usable form.

Two hundred million years ago the area of Three Gorges which Peng-Tzao was probably writing about, was a sea. After a series of geological upheavals, the Himalayas rose up, and the south-west area of China became land. The sea water trapped in the earth became condensed and gushed out as brine springs.

Archaeologists working in the area have found clear evidence of salt-making. Small red pottery cups with a pointed base and large round-bottomed pots have been discovered in such profusion that it clearly indicated the presence of a salt industry which used them for evaporation and transportation.

As you would expect, other ancient civilizations also have left behind references to salt, on clay tablets from Babylon and on papyri from Egyptians. Egyptian art from 1450 B.C. also records salt-making. At some of the old Egyptian burial sites from locations near the desert, human corpses have been found to be preserved with their flesh and skin intact. These aren't mummies; these are corpses which were preserved simply because the dry desert air contained salt which acted as a preservative.

India has a history of salt-making which is probably five thousand years old. There is a huge quantity of rock salt deposits in Punjab (now in Pakistan). On the west coast, the 23,000 sq km marshland of the Rann of Kutch is covered by sea water during monsoon months. The dry and hot winds evaporated the water, leaving the area covered with salt. The absence of fresh water discharge from rivers and low rainfall ensures that the salt is of good density and purity. In the east, there was a natural sea salt zone over 500 km long and fifteen to 100 km wide. The area was flooded by spring tides, saturating the soil with salt when the water later evaporated. Incidentally, the *Arthashastra* records that Chandragupta Maurya appointed a special official called *Lavanadhyaksha* to supervise the salt market and give out licences for salt manufacture. A separate sub-caste of salt diggers also came into being.

Ancient Greece had a far-flung trade based on the exchange of salt for slaves. The Jordanian town of Es-Salt, located on the road from Amman to Jerusalem, was known as Saltus in Byzantine times. In Austria, Celtic miners had dug several kilometres of galleries three hundred metres under a mountain some three thousand years ago. On the Atlantic coast, people collected salty sand and produced salt in clay pots by heating and evaporation two thousand years ago.

It is possible that civilization began on the edges of the desert because of natural surface deposits of salt found there. It is also likely that the first war was fought near the ancient city of Es-Salt.

In the China of 250 B.C., in a province which is now Sichuan, the governor was a man named Li Bing. According to Kurlansky, he must have been the world's first hydraulic engineer, certainly one of the most advanced, because he built a dam which is still in existence. He went on to build a series of dams which turned an area prone to cycles of droughts and floods to one which became agriculturally stable, and eventually, prosperous.

What concerns us here, though, is that he used his engineering skills for salt-making. Sichuan probably produced salt from as early as 3000 B.C. But Li Bing found a way of increasing its production by going to the source of the brine: he went beyond the pools from which it had been collected until then and developed drilling tools to make the world's first brine wells. He also pioneered the use of bamboo piping to get the brine out; in later years, many areas of China were using bamboo piping, not just for brine but even for plumbing. His work also led to the invention of percussion drilling, well before any other country used it.

It is obvious that the Chinese recognized the importance of salt: they were the first to tax it. Successive emperors either imposed a form of production control and then taxed it or merged smaller salt producing centres and merged them to make them more effective. Later emperors took over production completely to make salt a state monopoly whereby they could profit from its sale as well as from the tax they imposed on the consumer. These revenues were used to finance their armies or in times of war. Even the Great Wall of China was built out of tax revenues.

Salt taxes changed from time to time. They were lowered when there were signs of unrest or the leadership wanted popular support. They were also lowered when at a philosophical level they were found to be unjustifiable as at the end of the first century A.D. This about-turn followed the rule of the Tang Dynasty which had increased salt taxes to a level where they constituted half the state revenue. This had led to protests and even riots.

The practice of salting of goods may have originated either in Egypt or in China. There is evidence of the salting of fish in the Three Gorges area showing that this began somewhere in

2000 B.C., but it is likely that Egyptians salted fish and birds much earlier as these have been found in tombs to give the dead sustenance in the afterlife.

Egyptians even salted vegetables: the fruit of the olive tree, the olives in brine which we eat today, originated at that time. Cannily, the Egyptians realized that salt was too bulky to export, so they went into the world's first 'value added' export: they salted fish, vegetables and olives and traded them in and around 2800 B.C. for Phoenician cedar, glass and dye.

In other parts of Africa too salt played an important role, particularly in the trading empire of Mali. Although it is difficult to imagine, caravans of as many as 40,000 camels carried salt across the Sahara to Timbuktu, the 700 km journey taking a month or more. Due to its location the latter was the centre of trade for most of West Africa. Incidentally, there are stories told of an entire city built of salt called Taghaza. The reasons to use this unusual building material weren't aesthetic; salt blocks just happened to be the one building material conveniently available.

The one Egyptian use of salt everyone knows about is mummification. Egyptians believed that the spirit of the departed would return to the original body in the afterlife, which made it important to preserve the body. The process depended on desiccation because bacteria and fungi, like all other living organisms, need water to survive. Therefore, if water is removed, putrefaction will not occur.

Mummification was carried out in designated places called 'The House of the Dead'. It was work requiring the skill of surgeons, yet the workers became outcasts because they were feared to carry infections from the dead.

The process of mummification took as long as seventy days. Embalmers removed the brain through a nostril with a hook. Internal organs, except for the heart and kidneys (which were left in), were removed through an incision in the abdomen. The empty body was filled with linen pads, or sometimes just with sawdust. It was then placed in Natron salt until all tissues dried out. The body was wrapped in many layers of linen strips and placed in a coffin. The organs which had been removed were wrapped in a cloth with Natron salt and put in jars. These were placed alongside the body along with important objects belonging to the dead person for use in the afterlife. (Incidentally, Natron salt is a combination of sodium bicarbonate and sodium chloride which was obtained from the Natron Valley in the desert of Wadi El-Natron, and the word *mummy* comes from the Arabic *mumiya*, meaning embalmed body.)

The Romans too were salt savvy. Most Italian cities were founded near salt works, starting with Rome. The first great road built by the Romans was Via Salaria (Salt Road) to bring salt to Rome and the interior. Since they were prodigious builders of infrastructure, they also built salt works wherever they could, for example, in England. Romans were also more enlightened as rulers than other dynasties, and so did not set up government monopolies but instead controlled prices. In fact in 506 B.C., Rome's biggest salt works were temporarily taken over because the rulers felt that prices were too high.

Generally, the aristocracy subsidized the price of salt to make it affordable to the plebeian, the common man. (The phrase 'common salt' originates from here.) The importance of salt to plebeians can be gauged from this interesting little fact: when Emperor Augustus began preparations for a naval campaign to defeat Antony and Cleopatra, he tried to win over public opinion by distributing free olive oil and salt. The Roman Holiday for salt, though, ended during the Punic Wars (264 to 146 B.C.) when salt prices were raised to finance the war machine.

Romans used salt extensively to preserve food. Hams and sausages were widely served, while fish (tuna, mullet, mackerel and sardines) were the centrepiece of cuisine and trade. Some species of salted fish were also seen to have medicinal value.

Olives, of course, were consumed by everyone, but here there was a social difference: for Patricians (aristocracy), they were just starters; for Plebeians (commoners) they were the main course. Romans salted their greens too, and the word 'Salad' comes from *salada*, based on the Latin *sal* (salt).

In looking at the history of salt, it is amazing to note that nothing much changed from ancient times almost to the twentieth century: salt was a prized commodity; wars were fought over it; taxes were imposed on it; food was preserved by it ... Only the dates on the calendar were different.

Salted foods continued to be a necessary part of life everywhere. In medieval England, the annual slaughter of animals was around Martinmas, St Martin's Day (10 November). This meant that fresh meat in large quantities was available only around this time; it was later salted for use in winter. Fresh protein in the form of milk was then available only from the limited number of cows kept under shelter. This pattern was followed in most countries with local variations. In Sweden, for example, the slaughter took place earlier to keep pace with winter.

Beef and pork and meat were dried and salted and joints, hams and sausages. Butter was salted too (one kg of salt to ten kg of butter). Fish, freshwater or from the sea, was also dried and salted, while bread was hung up to dry. One statistic shows that of 102 kg of beef and pork consumed by a family, as much as 99 kg was salted. Butter was often kept for three or four years. Washing down so much salt required the drinking of a large amount of beer. Soldiers' rations in Sweden, for example, had to be raised from five pints of beer a day to eight pints a day and as much as eleven pints on Sunday! That could also suggest that the salt in food was a handy excuse for some serious recreational drinking.

Compare that with the diet of soldiers in the American Civil War. In the early 1860s, each Confederate soldier was provided these rations:

Starch consisting of 11.5 kg coarse meal
3 kg flour and biscuits, 1.5 kg rice
Protein consisting of 4.5 kg bacon
Salt, approx 0.7 kg

That's a lot of salt whatever your language (or your drawl). And when you consider that horses needed salt too, it made for a heavy requirement of salt altogether. This soon led to a great shortage of salt so that by 1862 only half of the meat in Georgia could be saved for the next season. And packets of salt were given as wedding presents!

Earlier, salt had played a prominent role in the European exploration of North America and therefore of American, Canadian and Mexican history. Major European fishing fleets had discovered the fishing riches of the Grand Banks of Newfoundland at the end of the fifteenth century. The Portuguese and Spanish fleets used the 'wet' method of salting fish, i.e., on board the ship, whereas the French and English fleets used the 'dry' method, i.e., on shore. If it hadn't been for the latter, the first inhabitants of North America may not have landed there, at least not for a while.

Salt taxes, as they had done in ancient times, continued to play a significant role in every country's politics. In the late eighteenth century in England, the British monarchy raised taxes to such an extent that a black market in salt became rampant. In 1785 alone, ten thousand people were jailed for smuggling. More recently when they reached exorbitant levels in China in the early twentieth century, they created discontent which resulted in the toppling of the imperial government.

The most far-reaching effects of salt taxation were felt in France. The French monarchy levied tax on salt of an increasingly complicated kind. Under Philip VI in the fourteenth century a salt tax administration, Pays de Grande Gabelle, was instituted. As the taxation increased, 40,000 farmers rose in futile armed rebellion in 1543. By the mid-seventeenth century under Louis XIV, salt tax had become the leading revenue earner for the monarchy. As if this weren't enough, every adult was *compulsorily* required to purchase 7 kg of salt per year at a price fixed by the government. Since this was more salt than humans could ordinarily consume, the only solution was to use it for salting foods. So the government passed a law prohibiting this!

The Gabelle was also revised to suit political ends, being reduced for certain provinces (like newly acquired ones) and reaching punitive levels in others. In 1784, according to Kurlansky, a Minot of salt (around fifty kg) cost thirty-one sous in Brittany, eighty-one sous in Poitou, 591 sous in Anjou, and 611 sous in Berry.

These differentials, obviously, were an invitation for a parallel black market in which smugglers became popular heroes and women were inducted into the trade, their corsets, bustiers and posteriors becoming hiding places for smuggled goods. Even salted fish was over-salted several times to smuggle salt! As many as three thousand men and women were jailed every year for salt smuggling.

This wasn't, of course, the only reason for the French Revolution, but it was certainly an important factor. When the revolutionary government took over, one of its early actions was to abolish salt tax and free all salt offenders.

The first step in reducing the importance of salt was taken around the time of the French Revolution. In 1795, Nicolas Appert a French chef began to process food in salted containers. In 1809 he won a prize offered by the government for a simple way of preserving food for the French army. His method was to use boiling water to heat food in glass bottles and then sealing them to make them air-tight. A year later in England, an English merchant named Peter Durand hit upon the idea of using cans rather than bottles, and patented the process. When Louis Pasteur discovered (1860) that organisms cause food spoilage and devised the method called pasteurization, it spelled the beginning of the end of salt's importance.

Around this time, commercial freezing of food had also started in the United States, but this was by a process of slow freezing. In 1925, a Massachusetts inventor called Clarence Birdseye developed a quick-freezing process. The difference between the two processes was vital: fast freezing produces small crystals of ice which do not interfere with the tissue structure of the frozen meat or vegetable so that it remains fairly close to its original state. Three years later, the Postum Company (later General Foods) which had bought Birdseye's patents, had already sold as much as one million pounds (453,600 kg) of frozen food.

Salted food was now no longer a necessity. The inevitable happened: salted food became a delicacy. Edible salt was no longer as important, besides which more modern methods of salt production meant that most countries could have their own supply. Whatever happened, one thing was certain: no government would now be able to support itself on salt revenues.

SALT WARS

Since salt was such a precious commodity, it is not surprising that many wars were fought over it. Some had to do with preserving a monopoly of production, some for dominating trade; in other cases, salt may not have been the main factor, but it played an important part in the battle. What follows is illustrative of the role salt, the condiment we dismissively call common salt, played in many uncommon conflicts.

CONFLICT ONE: GENOA v. VENICE

These were the superpowers of the Mediterranean area around the thirteenth and fourteenth centuries A.D. Salt could be produced on any seashore in the Mediterranean, but these two city-states exerted almost total control over the trade in their spheres of influence.

From the eighth century, Genoa had become a strong naval power. By the twelfth century, it had established trading settlements in Constantinople (now Istanbul), Cyrus, Syria and Tunis. In the 1200s, Genoa was at the height of its powers, controlling the central Mediterranean area including the islands of Corsica and Sardinia.

Venetian economy from the 400s to 800s was based on fishing and trading, with the bulk of the latter being on the Adriatic coast. During the ninth century, Venice expanded its operations and its trading partners included Constantinople, cities in the Italian mainland and the northern coast of Africa. By the 1200s, Venice too was powerful. Thus, its rise to power paralleled Genoa's and in their simultaneous ascension to power in the same geographical area, there was the potential of a decisive conflict.

Their general clash of interests was heightened by their specific attempts to dominate the salt trade. The Venetians were the aggressors here. While salt was one of many items for commerce for the Genoese, for Venetians it was the commodity. They had begun to produce it in the sixth century and when a rival salt power arose in Camacchio, they destroyed it in 932 A.D. By the end of the twelfth century, their salt powerhouse was Chioggia, controlled by the Doge of Venice, but another powerful personality, the Archbishop of Ravena had set up a rival power centre for salt in Cervia. Mutual trade embargoes resulted; in the end it was the Doge who was victorious.

Later the Venetians' salt strategy changed from controlling production to controlling trade. Venetian merchants bought, shipped and sold salt, all under the watchful eye of the state body set up for the purpose, the *Collegio del Sal*. Salt was bought at one ducat a ton while shipping cost three ducats a ton. After a complicated arrangement of state subsidies and taxes, the selling price had gone up to thirty-two ducats a ton! The saving grace of this profiteering was that it gave rise to some of the most elegant architecture and collections of art, still to be found today in the city of Venice.

In the meantime, Venetian and Genoese attempts at total control of trade, led to a series of wars in the 1300s, culminating in the defeat of Genoa in 1380. Venice's power, then, was such that it ruled Crete, Cyprus, the Dalmation coast and North-east Italy. Venetian ships carried all silks, spices and other goods from

Asia to Europe. No salt could move in the Adriatic either, unless it was carried in Venetian ships which passed through Venice.

CONFLICT TWO: THE SEVEN YEARS' WAR

The Seven Years' War took place between 1756 and 1763 and was principally between Austria and Prussia. Yet, by the time it had got seriously underway, it involved most nations in Europe, with Great Britain and Prussia on one side and France, Austria, Russia, Sweden and most German states on the other. For Britain, it was a proxy war fought by Frederick the Great of Prussia, largely on German soil. Britain extended aid, nominally at first, then wholeheartedly.

Britain got involved in the war as an extension of its conflict with France in North America which had broken out two years earlier in 1754. The cause of the conflict was the clash in the territorial ambitions of both sides in America. It also had to do with control of the fishing industry in the Atlantic and Britain's wish to do away with its dependence on French salt supplies, required, ironically, to preserve the Atlantic catch and for making gunpowder. British salt produced at Liverpool and Cheshire was not enough, besides which the latter was feeling the effects of both deforestation (for fuel required for salt making) and soil subsidence (due to excessive salt mining). Canadian and American salt, thus, would come in rather handy.

The French and Indian War, as it was called in America, ended along with the Seven Years' War in Europe, with a beaten France conceding almost all its land holdings in Canada and some of its land in America. It also ceded its control of territory in India to the British.

CONFLICT THREE: THE AMERICAN WAR OF INDEPENDENCE

The French and Indian War may have brought victory to Britain, but it was the forerunner of American independence:

the long wars on two fronts had so taxed the British treasury that its national debt had doubled. Besides, Britain had to now control enlarged territories in Canada, America and India. Under instructions from King George III, the British parliament passed laws which restricted the freedom of operation of British colonists in America and simultaneously raised taxes on them.

The reaction was fast and furious, and on 19 April 1775 a mere twelve years after the end of the French and Indian War, the Revolutionary War broke out. On 4 July 1776, the Americans declared their independence from their British rulers, although it took a full seven years more before the British were finally defeated and independence became a reality in 1783.

During the war, one of the British strategies was to cut off salt supplies to American forces. They cut off salt from Massachusetts, which had been a key supplier earlier. By the end of 1776, the British army controlled the area around what is now New York and served the link between New England salt and Pennsylvania. British forces also carried out search-and-destroy raids. In April 1778, for example, they mounted a covert operation along the New Jersey coast and destroyed a number of salt works which were supplying salt to the revolutionary forces. In addition, since the British navy controlled the seas, they could cut off salt shipments coming in to the revolutionary forces from overseas suppliers from the West Indies and Europe. No wonder salt became expensive with prices jumping from pre-war days to the end of 1777 by as much as seventy to 140 times.

The American side then went into emergency mode, Pennsylvania and New Jersey, in particular, encouraging private salt production by offering cash incentives and supplying anyone and everyone with instruction manuals for salt making. In New Jersey there was a new Gold Rush: it was now the Salt Rush. John Adams wrote, 'All the old women and young children are going down to the Jersey shore to make salt. Salt water is boiling all over the coast.'

By the middle of 1778, there were so many people in the salt business that margins became smaller and cash incentives were reduced. Simultaneously, the French, still smarting from their defeat in the French and Indian war, entered the fray on the side of the revolutionary forces. The French navy was particularly useful: it helped ease the salt blockade by confronting the Royal Navy so that salt from Europe and the West Indies, could be imported.

When independence finally came, everyone was in a celebratory mood. Everyone, that is, except the salt makers of New Jersey.

CONFLICT FOUR: THE AMERICAN CIVIL WAR

As the French-Indian War was a watershed for the Americas, so was the Mexican war which started in 1846 and ended two years later. In this war the United States defeated Mexico and gained from it land that is now California, Nevada and Utah, most of Arizona and parts of Colorado, New Mexico and Wyoming. Mexico also agreed to recognize Texas as part of the United States.

In an uncanny echo of what happened to Britain after the French and Indian War, famous victories do not necessarily result in prosperity; they can also bring about much grief. In the United States, the vast new areas which the new nation had acquired suddenly heightened tension between the north and the south over the slavery issue: Northerners wanted slavery to be outlawed in the new territories, a move Southerners opposed bitterly.

Things would probably have come to a head anyway: the dispute had been simmering far too long. America was split in two on the subject, and in January 1861, South Carolina, Alabama, Florida, Georgia, Louisiana and Mississippi seceded from the Union. Later, Arkansas, North Carolina, Tennessee, Texas and Virginia also seceded.

The Civil War to prevent the dissolution of the union of the United States began on 12 April 1861 and four years later, after a war which killed more Americans than any other battle in history, (360,000 Union troops plus 260,000 Confederate troops), the North won, the Union was preserved and slavery was abolished throughout the land.

One of the factors which led to the Southern defeat was the shortage of salt needed for a multiplicity of purposes: cooking, preservation of food, for the healing of wounds and for making gunpowder. General Sherman of the Union troops recognized this. 'It (salt) is as important as gunpowder,' he said. 'Without salt they cannot make bacon and salt beef... Without salt, armies cannot subsist.' (As an example from another part of the world, Napoleon's retreat from Russia resulted in thousands of deaths, which could have been prevented if there had been enough salt to heal wounds and increase resistance.)

Union forces were under instructions to destroy any salt works which supplied salt to Southern forces. They fought a 36-hour battle over Saltville, Virginia, in order to capture an important salt plant which was said to be essential to the South's supplies. There were a number of other attacks by gunboats on saltworks at Fort William, at Goose Creek, at Long Bar and a number of other picturesque names where salt-making was important. The incessant attacks and harrying of anyone involved in salt contraband resulted in a desperate situation for the south. So desperate that the Confederate President Jefferson Davis offered to waive military service to anyone who was willing to go to the coast and make salt for the war effort.

CONFLICT FIVE: THE BRITISH IN INDIA

> It is unlawful for an Indian to carry a pail of sea water to his home. Although India has four of the world's best rock salt areas, the British government dumps some 600,000 tons on the Indian market annually, thus provides ballast tonnage for British

shipping, gets 20,000,000 dollars annual revenue from India. The monopolized salt is sold to Indians at prices sometimes 2000 per cent of production cost. Indian farmers who take cattle to the seashore at night to let them lick whatever salt is deposited, thereby run the risk of imprisonment.

That quotation from 1930 isn't from an Indian political leader or a nationalist newspaper, it's from *Time* magazine, at a moment in India's freedom struggle when the American media had come to sneer, and stayed back to give to the Western world the first objective reports on the real state of British India.

The story goes back to mid-eighteenth century. Liverpool and Cheshire, England's main production centres for salt, had increased production and were aggressively looking for new markets. Bengal, under the control of the East India Company, was the obvious target.

As it happened, Bengal had always produced its own salt. To make Bengal Salt uncompetetive, the British began to impose taxes on it. At first, tax was imposed in the form of 'land rent' and 'transit charges'. Later, it was made into a comprehensive salt duty. The British began to import more and more salt from England. Imports, starting from 352,835 maunds in 1846-47 increased to 1,012,698 maunds in 1850-51 and 1,850,762 maunds in 1851-52.

The impediment to greater imports was Orissa salt, right next door to Bengal. It made both natural salt through solar evaporation (Kartach salt) and through boiling (Panga salt). The East India Company, in fact, had been a regular customer, buying Orissa salt for a variety of purposes, including the making of munitions and gunpowder.

In 1790, the company had tried to get into a contract to buy all Orissa salt. Orissa was then under the control of the Marathas, and its Governor, Raghuji Bhonsle, immediately saw through the ruse: this was a game-plan to make the industry dependent on the British, then cripple it. 'No thank you,' Bhonsle said.

The company then took the extraordinary step of banning the entry of Orissa salt into Bengal. This led to large-scale smuggling which, given the porous border between the states and the jungle terrain, was difficult to control. Orissa salt, now contraband, still outsold English salt in Bengal. So the British did what empire builders do: they annexed Orissa and attached it to Bengal. They could now control the salt trade. All salt had to be sold to the British at a price decided by the British. No salt could be transported. It was only a matter of time before the British took over the manufacture as well. A Salt Act made violation of the manufacturing clause punishable with confiscation of the produce as well as six months' imprisonment. Any salt revenue officer was entitled to forcibly enter any premises where salt was suspected to be manufactured and seize or destroy it. There was local resistance, and some of it took violent form in 1817 in which salt workers attacked British salts works, but that was subdued with a brutal hand. The East India Company could now do with Orissa salt whatever it wanted to.

What it did was to create a 'Customs Line' in which 'Customs Officers' could impose duty on salt crossing state borders. Over the years, this line grew and grew, till it was a massive 4000 km long and had 1720 guard posts. It started in Torbela, beyond Rawalpindi, followed the Indus River, went through Multan, skirted Delhi and Agra, went past Jhansi, Hoshangabad, Khandwa, Burhanpur, then via Chandrapur to end at Sonapur on the Mahanadi River in what is now Orissa.

What is even more extraordinary, is that gradually the British made this 'line' into a proper border, complete with a hedge. This Great Hedge of India is a well-kept secret unlike the Great Wall of China. According to annual reports of the India Inland Customs Department, during the 1860s and 1870s the Hedge 'is a live one, from ten to fourteen feet in height and six to twelve feet thick, composed of closely clipped thorny trees and shrubs.' These included babool, plum, carounda, prickly pear and thuer. A thorny creeper, (*Guilandina bondue*) was planted to intertwine with the shrubs and trees. The Hedge didn't go for the whole length of the Customs Line but for a considerable portion of it, starting from Leia, near Multan upto Burhanpur.

By 1877, the state of the Hedge was described as follows:

"Perfect" or "Good" condition:
- Green Hedge   411.50 miles
- Green & Dry Hedge   298.15 miles
- Dry & Hedge   471.75 miles
- Stone Wall   6.35 miles

"Wanting" or "Insufficient" condition   333.25 miles

TOTAL   **1,521.00 miles**

This barrier was manned by a virtual army: 136 officers, 2499 petty officers and 11,288 men, making a total of 13,923 people. And this number was growing. The British were truly waging an epic Salt War against their Indian colony.

If anyone had any illusions of the 'benign' nature of the British Empire in India, they only needed to look at the way India's salt industry was first repressed, then destroyed, so that until its revival in independent India, the country had become completely dependent on imported salt.

DEAD SEA, LIVE SEAS

No one has actually walked on the Dead Sea, though some have claimed to sit on it. Everyone though can float on it. In fact, no one, human, cow, elephant or camel, can sink in it.

The Dead Sea is the lowest place on the surface of the earth, four hundred metres (1312 feet) below sea level. It is located between Israel and Jordan and, in fact, forms part of the border between the two countries. It gets its name from one simple fact: nothing, almost nothing except for some plants and brine shrimp can live in it. This is due to its contents being the saltiest body of water in the world, so salty, that it is nine times as salty as any ocean. This gives it the extra buoyancy which enables a person to lie on his back, read a book, smoke a cigar or doze off without sinking. The saltiness, in fact, behaves like a waterbed or like an air-filled flotation mattress.

The Dead Sea is actually a lake, completely landlocked, surrounded by rocky and barren lands, as well as steep and brightly coloured cliffs on two sides. It is a large lake, though, because it covers over a thousand square kilometres. It is 80 km long and, at its widest, 18 km wide. The lake's deepest point is 400 metres.

When people began to call the lake Dead Sea, they didn't

know how apt the name would be, because the Dead Sea is dying. From early in the twentieth century, its water level has been falling at the rate of thirteen inches per year. On a geological time scale that's like an express train. This is probably due to a combination of the limited rainfall in the area (100 mm annually, i.e., four inches) and the very high temperatures which cause rapid evaporation. Freshwater comes in from the River Jordan and other rivers and streams, making the Dead Sea's uppermost surface less saline than lower layers. Forty metres below the surface the salinity has increased, going up to 300 gm of salt per kg of water. Below that, the water reaches saturation levels. The water is surprisingly clear while the shore is brilliantly white due to the crystals of salt which cover everything.

The Dead Sea's excessive saltiness has been ascribed to a combination of factors. Primarily, while the River Jordan and the smaller rivers and streams flow into it, there's no river flowing out. Thus a variety of minerals and salts come into the Dead Sea (sodium chloride, bromine, calcium chloride, potassium chloride) while none of them go out. What goes out is only through evaporation, which is water. And a lot of water goes out due to the extreme heat in the area, so that evaporation is continuous and heavy. What you get left behind is a concentration of salts and minerals. Another question, relevant to this context is why the Dead Sea is like a lake, cut off from other waterbodies. The explanation for that can only come in geological terms and in geological numbers: the Dead Sea was likely to have been formed some five million years ago when the Arabian Peninsula and the African continent shifted and formed the Great Rift Valley. Before that, it is said that the Dead Sea was connected to the Mediterranean near what is now Haifa port. The geological shift pushed the Galilee Heights to their current level and these newly formed mountains cut off the Mediterranean from the Dead Sea.

The Dead Sea is mentioned in the Bible as the Salt Sea. The Old Testament carries the story of Lot's wife who was turned into a pillar of salt because she disobeyed God's injunction not to look back at the destruction of Sodom and Gomorrah.

No one quite knows where Sodom and Gomorrah were geographically. What everyone does know is that the shores of the Dead Sea contain a number of salt columns, and visitors are told that one of them is Lot's wife. Which one of the columns, is never specified, which is convenient, especially since these columns do drop off occasionally. Can anyone prove that a column was Lot's wife? Obviously not. But then, no one can disprove this either. That's useful for the tourism industry. (Incidentally, the Lot family's strange fate does not end with the wife becoming a pillar of salt: according to the Old Testament, Lot and his two daughters did escape the destruction. In later years, the daughters despaired of getting married and having children. So they got their father drunk and had sex with him. Each had a child, as a result. The two sons Moab and Ben-ammi are considered the ancestors of two tribes living on the East Bank of the River Jordan.

The fecundity of saltwater is seen even in the Dead Sea. It may not support any life to speak of, but it is rich in minerals. Natural asphalt from it was used for caulking ships in Roman times. Now two companies, the Dead Sea Works and the Arab Potash Company draw salts of various kinds as well as minerals from it and sell them commercially. In addition, several spas have sprung up in the area in the belief that the Dead Sea water has health benefits. Useful, again, for the tourism industry.

The fecundity of saltwater oceans is so large, that human beings have only skimmed the surface, so to speak. Fishing, of course, is the industry most nations have exploited, perhaps over exploited, but in the future we might get as much as five times the food we now get from the ocean. And there is much more that oceans will provide us in the future. There's water, and as we find present sources inadequate, at a future date, we will have developed a cheaper way of removing salt from sea water. The first tidal power plant in France, uses the force of the

The foundation stone ceremony for Okhla Salt Works, 1929 (precursor to Tata Chemicals). The ceremony was performed by H.E. Diwan Sahib of Baroda on 4 May 1927. Seated in the car are from left to right: H.H. Shri Lalsingh Maharaja (brother of H.H. Sayaji Rao, Maharaja of Baroda State) H.E. V T Krishnamachariar (Diwan Sahib of Baroda State) and Shri Pilaji Rao Bahadur. Although this area was not contiguous with the Baroda State, for effective administration, required due to the martial tribes that inhabited the region, it was handed over to the Baroda State by the British.

Photo courtesy: Vivek Talwar

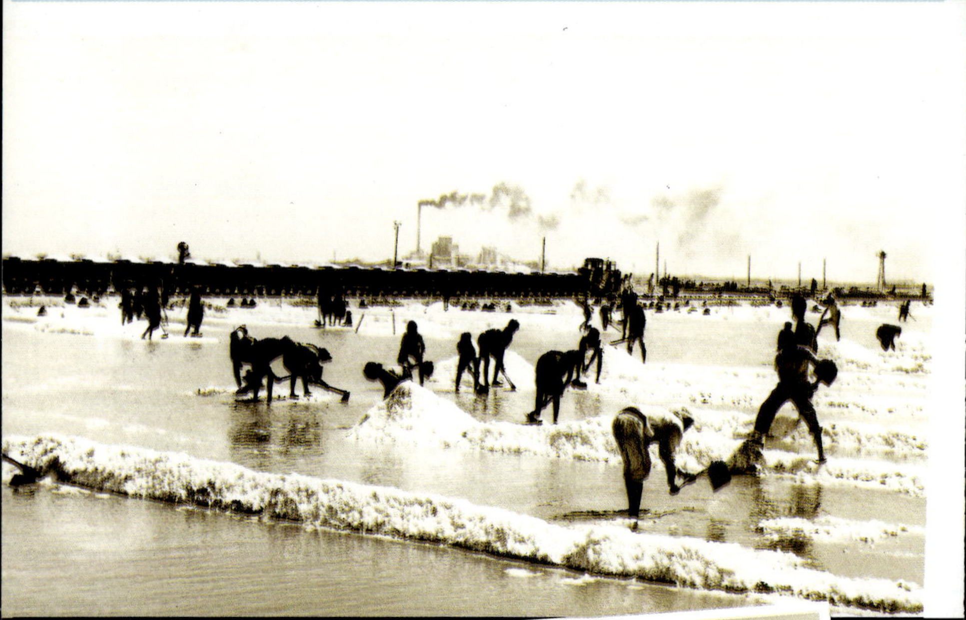

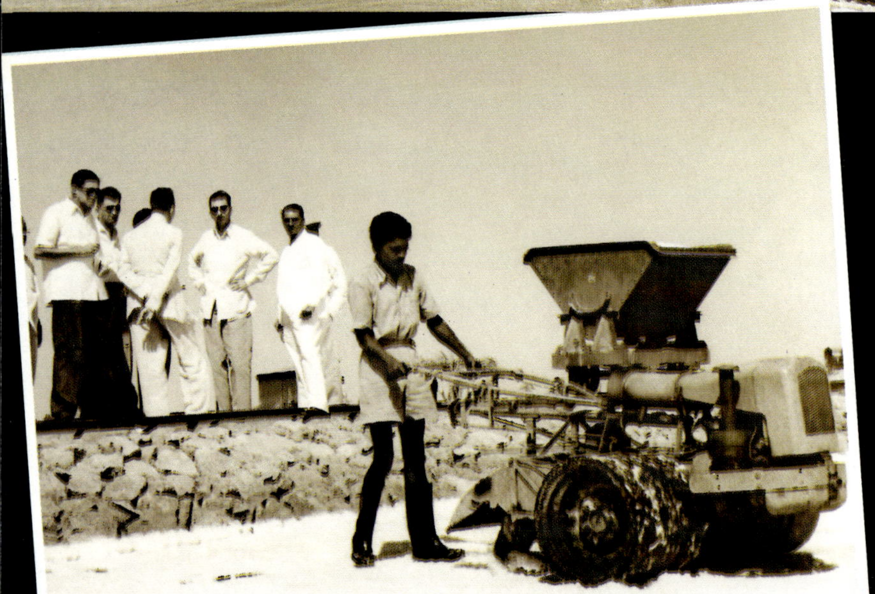

ABOVE:
Pillars of Mithapur: The men who run the salt plant.

FACING PAGE (TOP):
Harvesting salt, 1950: Mithapur provides ideal conditions for a very efficient salt works: high salinity in sea water (there are no freshwater systems draining into the sea for hundreds of miles on either side), flat expanse of land, high wind velocities, and low rainfall.

FACING PAGE (BOTTOM):
Parikh's invention, 1962: In-house innovation led to an officer of Tata Chemicals, R.A. Parikh, (extreme right) designing a mechanical salt harvester, which did the work much faster than manual harvesting. However, concerned about the livelihood of the local salt lifters, Tata Chemicals did not deploy the machine. J.R.D. Tata (second from right) inspecting the machine.

Photos courtesy: Vivek Talwar

ABOVE:
Rabari women displaying traditional handicrafts that Tata Chemicals Society for Rural Development helps promote as an additional livelihood activity.

FACING PAGE (TOP):
Rabari girls, who actively participate in the many community development activities that Tata Chemicals Society for Rural Development implements, including micro credit, handicrafts development, watershed development, animal husbandry, etc.

FACING PAGE (BOTTOM):
The Tata Chemicals' mobile clinic provides medical services to the forty-two villages of Okhamandal Taluka under its medical outreach programme.
Photos courtesy: Vivek Talwar

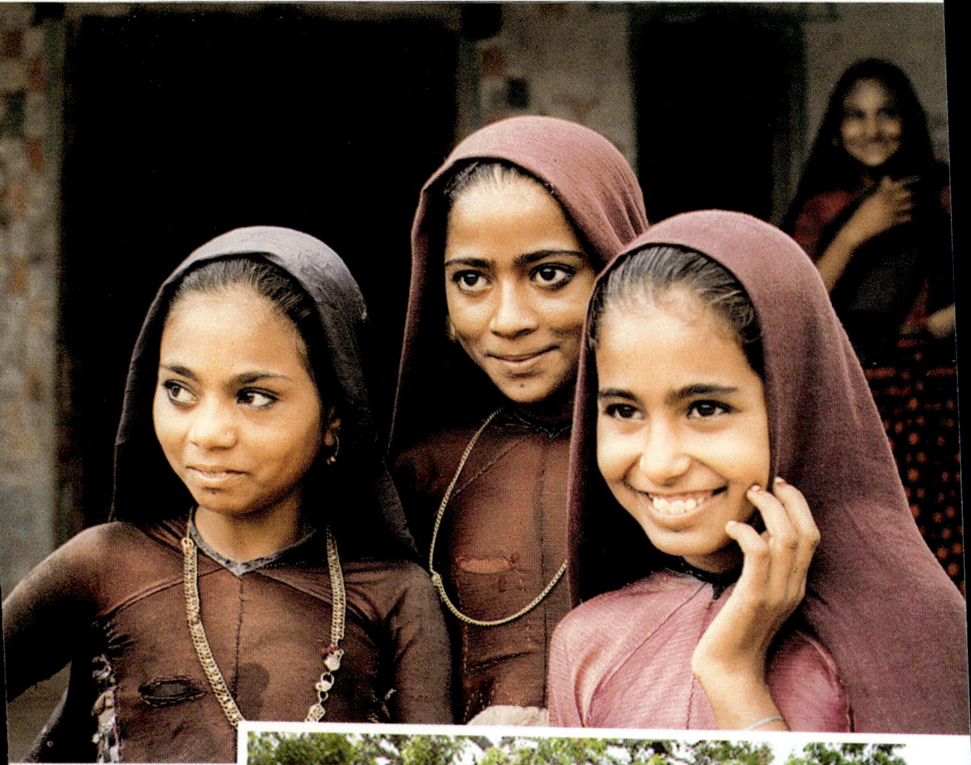

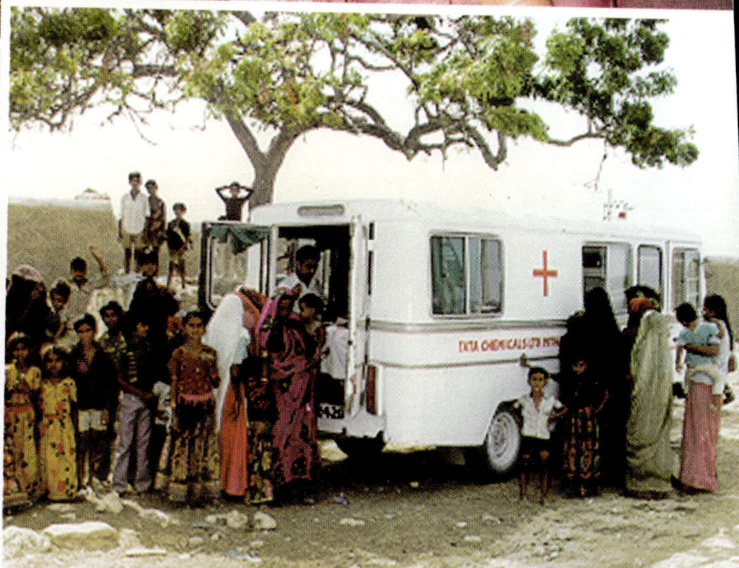

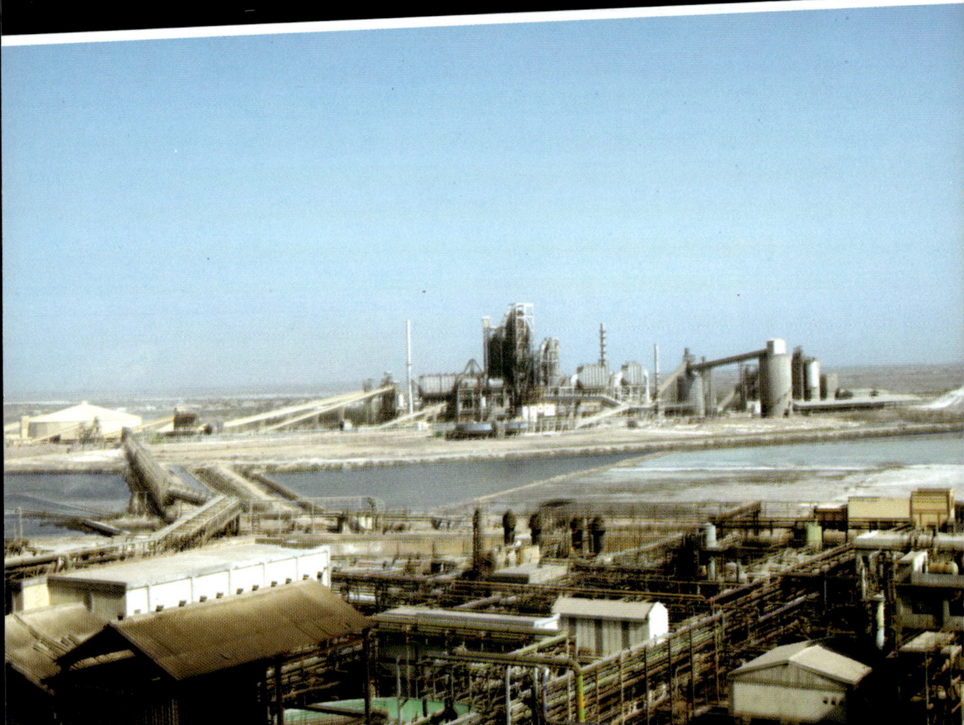

ABOVE:
A salt stack in the foreground, with a part of the cement plant in the background.

FACING PAGE:
A section of the salt pans separating the chemical complex (part visible in the foreground) and the cement plant in the background.
Photos courtesy: Govind Art Studio

ABOVE:
Flamingos in the salt works.

LEFT:
A Pelican taking off from the salt pans.
Photos courtesy: Vivek Talwar

tides to produce electricity. This will surely be of use in a future when oil wells begun to dry up. Then there are minerals, a seemingly endless supply of them.

When minerals on earth begin to get scarce, it is the oceans which will provide us with what we need: after all, sea water contains all the elements in the earth's crust. The composition of sea water is normally the same in all oceans. It goes like this.

Chlorine	55.2 per cent
Sodium	30.5 per cent
Magnesium	3.7 per cent
Sulphur	2.5 per cent
Calcium	1.2 per cent
Potassium	1.1 per cent
Others	5.8 per cent

No one has satisfactorily answered the very basic question of why the sea is salty to start with. Nor has anyone answered the related question of how the salinity of oceans is maintained more or less at a constant level year after year and season after season and has remained so for 1.5 billion years. This is surprising because, according to scientists, the amounts of minerals and salts added to the ocean from rivers, the weathering of rocks, from hydrothermal vents in the ocean and from material from volcanoes (both on earth and the sea), far exceeds the levels present in the oceans.

Calculations show that three billion tons of salt are added each year by these four principal means. How are they removed? There's evaporation from the water and its deposition on land. Then there are biological cycles using planktons, reefs, etc. Salt traps beneath continental shelves remove some quantity. Then hydrothermal vents have chemical reactions which might remove some quantity of salt. Lastly there's absorption and sedimentation by clay and other inorganic compounds. (Incidentally, there is a massive amount of salt deposited in the deeper parts of the Mediterranean, some of it as much as three hundred metres thick.)

According to James Lovelock, coral reefs represent an organic evaporation pond which removes salt from sea water. He contends that the Great Barrier Reef, off the coast of Australia, is a partly finished massive evaporation lagoon, and that marine organisms, through biological processes are able to remove salt in very specific proportions.

These unresolved questions point to a miracle that occurs continuously in our world's oceans. Add to that the fact that the composition of sea water and the composition of body fluids in man and animals show a remarkable similarity: the proportion of elements is virtually the same. 'Dust to dust, ashes to ashes,' somehow seems inappropriate. It is from our oceans that all life began ... And to the oceans we will return?

SUPERSTITION AND MYTH

In the beginning, there was only water. *Oos samay*, the *Vishnu Puran* tell us

> *Na din tha, na raat thi*
> *Na akash tha, na prithvi thi*
> *Na andhakar tha, na prakash tha*
> *Na inke atirikt kuch aur hi tha*

> (It was neither day, nor was it night
> There was no sky, nor was there earth
> It wasn't dark, nor was it light
> Besides these, there wasn't anything else at all).

Then Brahma took the form of Varaha, an amphibian half animal, half fish, and entered the waters. Earth saw Brahma in his avatar and said, 'Rescue me'. So Brahma did and took the earth out of the oceans.

Creation myths exist in all cultures as you would expect. They seek to explain the eternal mysteries of how the world was formed. What causes (or caused) universal catastrophes? What is death and is there an afterlife?

Cosmogony – myths of the birth of the world, generally do start with water. The Babylonians, for example, believed that in the beginning there was only a watery void in which freshwaters mingled with the saltwaters of the sea. The former was personified by Apsu, a male king, the latter by Tiamat, a female. The story goes on to describe a conflict between these earliest gods and a younger generation descended from them. Marduk, a new generation god of thunder and lightning, leads his army against Tiamat's (shown as a dragon) and defeats her. He splits her carcass in two to form heaven and earth and establishes the sun, moon and constellations.

Hindu mythology has another ocean-centred story, the story of *Amrit Manthan*. Under constant attacks from the *danavas*, it becomes imperative for the gods to get *amrit* out of the ocean to guarantee them eternal life. They utilize the services of Vasuka *naag*, the large snake, as the rope for churning the ocean and Mandrachal *parbat*, the famous hill, as the rod for churning.

The *devas* cannot manage on their own, while the *danavas* want the *amrit* too, so in a rare show of cooperation, the forces of good and evil begin the act of churning together. Vishnu transforms himself into a large tortoise which becomes a platform to stand on. The *devas*, cannily, have got the right end of the *naag*, i.e., the tail, while the *danavas* have the mouth. Before too long, the snake's venom has begun to kill the *danavas*. Shiva comes to the rescue and drinks up the poison (which is why his neck is blue and he is often referred to as Neelkanth).

The churning of the ocean brings out, by turn, Kamdenu, the cow; Kalpavriksha, the tree; Iyrawati, the elephant and many other elements which provide knowledge, medicine and jewels. It also brings out Lakshmi, before finally producing the *amrit*. The *danavas*, in spite of having the wrong end of the snake, snatch the *amrit* and make off with it. Vishnu comes to the rescue again, takes the form of Mohini, an incredibly beautiful woman who distracts the *danavas* long enough for the *amrit* to be reclaimed by the gods.

(Incidentally, drops of *amrit* are said to have fallen on a number of places: Nashik, Ujjain, Hardwar, Amritsar and Allahabad, thus sanctifying these sites to hold kumbh melas at fixed periods).

God, or gods, or gods and devils, may be in the details, but it is interesting to see that even in 2000 B.C., the Big Picture was clear and the ocean and its saltwaters were already being seen as the origin of our world and all that it contains now, something modern science has begun to look at seriously fairly recently.

The importance of an object is directly proportional to the myths and superstitions that grow around it. This is certainly true of salt. We were brought up with this discomfiting little superstition: if you spill salt and don't pick it up, it all goes into a ledger against your name. When your time comes, Yama, the God of Death, first brings out the ledger, then arranges a pile of the exact quantity of salt you have wasted in your lifetime and makes you pick it up. With your eyelashes.

Here's an echo of that from medieval England. There's the same ledger and the same adding up of a precious substance so wantonly wasted. And the comeuppance with a slight difference: here you don't need to flutter your eyelashes; just wait outside the gates of Paradise for as many years as the grains of salt you spilled in your lifetime.

And this one from Norway works on the same principle, though retribution is in this lifetime: you will shed as many tears as may be required to dissolve the salt you have spilled.

Salt superstitions follow a similar pattern, whatever the country, the culture or the period in time. These superstitions can be separated by their type. First there's the belief in the purifying and protective ability of salt. And where is this needed most but in newborn babies?

> The bairnie she swyl'd in linen so fine.
> In a gilded casket she laid it syne,

Mickle sault and light she laid there in,
Cause yet in God's house it had'na been.

This is an ancient ballad, *The King's Daughter*. It alludes to the dangerous time between the time a baby is born and can be baptized. To protect it, the mother puts salt and candles ('sault and light') by the baby's side in its 'gilded casket'.

These precautions weren't just for royal babies. Here's a little metrical statement from the early sixteenth century: 'It was uncrisned, seeming out of doubt, for salt was bound at its neck in a linen clout.'

In Sicily – better known as training ground for a certain trade practised in the USA and in Italy – they go, inevitably, one step further. The priest places a little salt in the child's mouth at baptism. That, apparently, imparts wisdom. So when a child grows up to be a dullard, people say, 'The priest put too little salt in his mouth.'

The need for salt as a means of purification (or protection) was obviously not restricted to babies and children. Adults needed it too. In 1789, when the Scottish poet Robert Burns wanted to move to a new house at Ellisland, he was escorted to it in a procession by his relatives. Pride of place at the head of the column was given to a bowl of salt carried on top of the family Bible. Burns wasn't being particularly superstitious; salt for a new home is a commonly-observed custom, and not just in Scotland. In Russia, for example, salt and bread are the first articles to be carried into the house of a newly-married couple. Before that, at the wedding breakfast, a servant will have carried a plate containing salt from table to table, so that guests can place presents of money in it. The wedding breakfast bit may be a Russian innovation, but salt-in-a-new-house is a tradition in many other countries, as different as Hungary and Wales. In Egypt, salt is used in a different kind of commencement ceremony: before a caravan sets out, women will throw salt on burning coals, which are then carried in earthen bowls and distributed throughout the caravan. 'May you be blessed in going and coming,' they chant in unison as they place the bowls reverentially.

A second category of salt superstition lies in its supposed ability to ward off evil spirits. In Hungary, for example, a housewife will sprinkle it on a regular basis on her threshold to protect it from witches and other evil spirits. So will a Japanese housewife (who will also ensure that she doesn't buy the salt at night because that reduces its efficacy).

Other local variations go like this: in Morocco, salt is worn in an amulet to ward off evil; the Neopolitan will suspend a piece of rock salt from his neck. In the mountain regions of Germany, three grains of salt in a milk pot will keep witches away from it, essential when the milk is to be given to children. They do that in Normandy too, while in Scotland the salt goes on the lid, a useful difference considering that salt in milk is likely to curdle it (then even witches won't want it). Whatever the variations, salt is considered an effective deterrent to witches and others, because these bad spirits are compelled by the rules of the game to count every grain of salt encountered by them, and that could take a long time.

That's probably the basis of the very common custom in many countries to throw salt over your left shoulder. Why over the left? Because evil spirits were generally reckoned to be on the left. After all, the word 'sinister' comes from the Old French word 'sinistre' or the Latin 'sinister', both meaning 'left' (This, incidentally, was the basis of the prejudice against left-handedness, common all over the world till modern times).

For those who wanted extra protection for their newborn, the salt ammunition was augmented with other weapons. In poor rural communities in Italy and Spain, for example, a light is left on in the baby's room constantly. In addition, an image of the family saint is fastened on the front door along with a rosary and an unravelled napkin. Behind the door is kept a jug of salt and a broom. If a witch comes, she will go away on seeing the picture of the saint. If she doesn't for some reason, the salt, the napkin and the broom will provide more than adequate protection because the witch will first have to count the grains of salt, then the unravelled threads of the napkin and finally the twigs of which the broom is made. That's a lot of counting for the witch while the household can count on dawn breaking before the witch is through, so off she'll have to go.

During the Middle Ages, it was also commonly believed that witches could not eat anything that was salted. From this arose the torture-method of interrogation of suspected witches, force-feeding them heavily salted food while denying them water. The inquisitors, incidentally, were advised to wear around their necks amulets that contained salt which had been consecrated on Palm Sunday.

Harry Middleton Hyatt did some seminal work on African-American hoodoo practice and published his findings in *Hoodoo, Conjuration, Witchcraft, Rootwork*, a five-volume, 4766 page work. There he describes many Protection Spells which use salt. Like this one:

SALT AND SALTPETER BATH FOR UNDOING TRICKS

> 1457. A person dat been tricked in de skin it's something dat is buried for 'em or laid down on de steps for ' em – de house been dressed. You take nine teaspoonful of cooking salt, you take one dime of saltpeter, use dat, and eight quarts of water, hot water – just like water for a bath. You pull off all of your clothes, ever'thing you got on, you get in there and take a bath in dat same water nine times.
>
> Dat same water – don't throw dat water away, you keep it in something like you take a bath in. Never rub upwards – always rub from here down.
>
> From there down. A person whut's been tricked in de skin, rub from here down and use dat water nine times, and de last

time you use dat water, take it and throw it towards de sunrise, soon in the morning before the sun rise, so you get rid of dat complaint.

Spell-casters, whether practising African-American hoodoo tradition or in the European folk-magic tradition, have one thing in common: they start with salt, a pinch placed in each corner of the room before they begin their spell. When the spell is meant to be purely protective, salt may be used by itself or, occasionally, combined with saltpetre and black pepper. For stronger spells, such as attacking ones to be used against enemies, salt is combined with red pepper and sulphur.

A third general use of salt is as a good-luck charm. In the south of England, lonely maidens pining for their lovers try the salt spell: They throw a little salt into the fire on three successive Friday nights. Simultaneously, they say these words:

> It is not this salt I wish to burn,
> It is my lover's heart to turn;
> That he may neither rest nor happy be,
> Until he comes and speaks to me.

If the spell works, the lover will appear on the third Friday. The opposite of this, to remove an unwanted guest, also needs salt: it can be sprinkled on the chair of the unwanted guest, or thrown behind them. Scottish fishermen have a traditional custom of 'salting' their nets for luck before they set off for their catch. They also sometimes throw a little salt in the sea to 'blind the fairies'. Similarly in Japan, Sumo wrestlers throw a little salt into the ring before entering it.

Last, and certainly not the least, there are a whole lot of superstitions about salt from all over the world. Like:

- In Italy, you don't offer salt to the wife of a friend. It's akin to making a pass at her.

- If an adolescent girl lays the table, but forgets to get out the salt cellar, it's a sign she is not a virgin.
- Salt is also used for divination: lay a pile on the table before going to sleep. If in the morning the pile is exactly as it was the previous night, the future is good. On the other hand, if the pile is disturbed, it's not a good sign at all.
- Spilling of salt, or the overturning of the cellar at table is, of course, a catastrophe quickly to be addressed. Romans exclaimed, 'May the gods avert the omen!' In seventeenth century France, they cancelled out the bad luck by immediately spilling some wine, which was a sign of good luck.
- Passing the salt to your neighbour at a table has always been tricky because according to some superstitions, you will soon pick a quarrel with him. Etiquette books dealt with this problem at some length. One solution was not to hand over the salt shaker directly, but to move it on the table within their reach. Another was to take some salt on the flat blade of a knife and with that, transfer it to their plate.

Many of these superstitions have disappeared over the years. Blame the Greek philosophers of the sixth century for that. They were the ones who began to debunk mythmaking, stressing instead the need for rational thought.

The monotheism of the world's major religions and their dismissal of anything outside their holy books as heresy also discouraged superstition and myth. Even Hinduism, with its large pantheon of deities and the opportunity that gives for multiple myth-making has now shifted to an increasingly rational view of the world. No wonder the power of traditional stories has begun to wane. We now no longer believe in them; we only see them as fey leftovers of an old and bygone world. Pity, in a way. Because with them has gone out much of the magic from our lives.

THE WAY OF ALL FLESH

The consumption of salt has been declining rapidly in the last century: the European of the twentieth century, for example, used half the salt used by the European of the nineteenth. This was due to the discovery of refrigeration and cold storage so that salt was no longer needed for the preservation of food.

Salt's fall from its pre-eminent position as the saviour of mankind was further hastened by the medical discovery that oversalting of food results in hypertension and high blood pressure and can cause cardiovascular problems. (Incidentally, a Chinese medical book had talked about this in some detail in the first century. But in the complete absence of an alternative food preserving agent, the book's warning was ignored.)

Predictably, the opposite view also has its advocates: a low-salt diet has been deemed unhealthy and can cause dangerously low blood pressure. And, ironically, heart attacks. More recently, it has been observed that athletes who died during a marathon had lost the delicate balance of water and salt the body needs to maintain; by drinking too much water, they had lost too much salt.

These findings, published in the *American Archives of Internal Medicine*, were corroborated by findings of the US armed forces in early 2002. Three young soldiers, a 19-year-old air force recruit, a 20-year-old woman trainee in the army and a 19-year-old marine all died during exercises which involved either walking or marching over long distances. The woman died while trying to produce a urine sample.

The common factor in all cases turned out to be that each had 'overhydrated'. Apparently, the body cannot excrete an excess of fluid quickly enough. The water then goes to the bowel, which takes away salt from the body, thus diluting the concentration of salt in tissues. This causes an imbalance of body fluids which can result in a fatal swelling of the brain.

Overall, medical advice would suggest a commonsensical middle path, stressing the body's need for salt in moderation. How much is that? We lose between 4 and 6 grams of salt in twenty-four hours. To replace and augment that suitably, 5 to 10 grams need to be consumed in the same period, depending on the level of activity, weather conditions and diet: a manual labourer pushing a heavily-loaded cart uphill at 12 noon in May in Delhi will obviously need a much higher level of salt and water intake than a well-fed office employee sitting in an air-conditioned building in Mumbai.

Luckily, unlike our ancestors, we now have a choice on the subject of salt intake in our food. We eat ham because we like it, not because that's the only way to preserve meat. Some even claim that it is an excellent source of protein and is also high in thiamin, iron and other vitamins and minerals. No one's yet made health claims for bacon, but it is on most breakfast menus in the West, an example of highly-salted meat which is likely to survive into the future for reasons purely of taste.

Other salted foods which will survive because they have become delicacies rather than necessities are corned beef, caviar, anchovies, canned sardines and herring, salami, sauerkraut, cheese and butter.

Butter is the prime example of a food which once had to be salted to preserve it, but now doesn't, yet still is because that's the way we like it. Traditionally, butter made by churning the milk of buffaloes was made in India as early as 2000 B.C. But it couldn't be transported and it couldn't be kept, so its consumption was local and rapid. Salting improved its shelf life, but the next big development in the manufacture of butter came only in the mid-1800s when Louis Pasteur introduced pasteurization to the world: by heating food (including milk) at specific temperatures and for specified periods of time, bacteria and other harmful organisms were killed so that the food wouldn't spoil. The other advance came in 1879 when Carl Gustaf de Laval, a Swedish engineer, patented a device called Centrifugal Separator to remove cream from milk at great speed, thus greatly increasing production.

Cheese is yet another example of a produce which has changed character. Nomadic tribes in Asia made cheese four thousand years ago, primarily to obtain a version of milk which would keep from spoiling. The invention of the pasteurization process meant that salt was no longer an absolute necessity for cheese, but can you imagine unsalted cheese? Yet, today, there are over four hundred kinds of cheeses made in the world, with over two thousand names (because some similar cheeses have multiple names). These range from soft (Brie, Camembert, Cottage, Cream) to semi-soft (Munster, Roquefort, Limburger, Port du Salut) to hard (Cheddar, Edam, Gruyere, Swiss) to very hard (Asiago, Parmesan, Romano, Sapsago).

Ketchup is an example of a salt-based product invented for preserving food which has now moved away from salt, from preservation and even from its main ingredients. Indonesians had a sauce they pronounced Kaychup, which was essentially a fish and soy sauce. It travelled to eighteenth century England and became an anchovy sauce which used vinegar and white wine. Americans, with a large tomato produce, turned it into tomato ketchup (often called Catsup). It took many years before salt became a minor ingredient of ketchup.

Pickles are yet another example of a necessity becoming a relish. The Chinese (of course) thought of it first, pickling vegetables to stop them rotting. They realized, fairly early on, that salt is essential to the process. Just over one per cent of the vegetable's weight in salt prevents rotting by stopping the build-up of yeast until sugars from the vegetables break down and produce lactic acid which acts as a preservative. Generally the fruit or vegetables to be preserved are first soaked in brine and vinegar before they are flavoured with seasoning.

The Indian coastal variation of the pickle is even more economical. It works on the principle that every large catch contains a quantity of smaller fish which will not have a market as fresh fish, so it is best preserved. The same goes for larger fish which don't find a customer. The drying process uses a combination of sun and salt from sea water. *Bombil* (Bombay duck) are dried in the sun till they shrivel into sticks called *kadya*. Medium-size prawns, after being cleaned, are left to dry till they become *soday*. Small shrimp when dried are called *sukat*. Mackerel and other larger fish are strung together like a necklace and hung out to dry between two trees. The result, in all cases, is a very dry, hard and chewy product with a strong flavour and an extremely powerful smell. At best, this is a developed taste, and the people who develop a taste for *sukat, soday, kadya* generally can't afford to buy fresh fish. Since they also do not possess refrigerators or have access to cold storage, they pickle vegetables and fruit as well. The general method is to chop the ingredient into small pieces, soak them in a lemon and salt juice for two hours, then let them dry in the sun. This is done with ginger, with raw mango, with ripe mango, cauliflower, *bhindi* (ladies fingers), bitter gourd (*karela*) and other vegetables.

Pickling is also used to ensure that nothing is wasted. Small, raw mangoes (*kairis*) which have fallen off trees would otherwise rot. The seed is taken out, cut into small pieces, salted and dried. This is a variation on an after-dinner mint. The *kairi* flesh itself is salted, dried and pickled. Tamarind is salted, made into balls and dried in the sun. The salt will prevent decay and keep away pests. Peas can also be made to last a year by being boiled in salt water, then dried.

What follows are recipes from Vasumati Dharker's yet to be published cookbook. These are unusual pickles: to start with, they are non-vegetarian. And they can well become the main course.

MUTTON PICKLE

500 gm	mutton
100 gm	ginger
500 gm	green chillies
250 gm	cloves of garlic
Pinch of mustard seeds (rai)	
$1/2$ teaspoon	hing powder (asafoetida)
4 teaspoon	chilly powder
2 teaspoon	haldi (turmeric)
$2^{1}/_{2}$ teaspoon	salt
300 gm	lemon
150 gm	oil

Wash the meat, apply a little haldi, chilly powder and 15 garlic cloves and ginger paste to the meat. Cook the meat in water till it's tender. Now fry the meat in a pan with some oil. Put aside the fried meat pieces. In the same pan, fry green chillies, ginger and garlic (both cut into long pieces). Take out the oil and put it in a saucepan. When the oil is hot, add hing, rai, chilli powder, salt, haldi, green chillies, ginger, garlic cloves, and the fried meat. Stir. Finally, add lemon juice or vinegar.

FISH OR PRAWN PICKLE

Clean fish or prawns thoroughly. Cut into pieces. After washing again, apply salt, haldi and lemon juice and then fry. Now follow the meat pickle recipe.

KHARA GHOSH (SALTED MUTTON)

1 kg	meat
3 teaspoon	salt
4 tablespoon	ghee (clarified butter)

Wash the meat than apply salt. Keep it for two hours. Heat ghee in saucepan, then add mutton pieces. Stir for a while. Now add water in meat. When the water starts boiling, turn gas low, and cover saucepan. Allow the meat to cook properly and the water to dry off completely.

Those were Maharashtrian recipes. Parsis, most of who live in Maharashtra, have quite a distinctive cuisine that includes the following salt-based recipes:

DRIED FISH

Fish pieces covered with salt are wrapped in leaves and kept for four days. On day five the leaves are opened and the fish is kept to dry. This dried fish can then be eaten for weeks. Or at least as long as the accompanying whisky lasts.

LIMBU NU ACHAR (LIME PICKLE)

Pieces of lime marinated in salt, chilly powder and a little sugar (to cut the bitterness) is dried and bottled as a pickle. Stays for years. Given to people when sick and have lost their appetite.

AVLA NU ACHAR (AVLA PICKLE)

Avla with salt is dried and preserved. Given to people who have jaundice.

The following recipe, in its simplicity, is reminiscent of the original recipe from Iceland for Salted Cod. We could be back a few centuries, except that we time-travel carrying an oven:

SALTED-COD PUDDING

500 gm	salted cod
60 gm	rice
3 ml	milk
3	eggs
Butter	

Preheat oven to 200C.

Boil the salted cod until fully cooked. Take it out of the water and leave to cool.

Skin the fish, debone it and make a stew out of it in a bowl, not adding anything.

Boil the rice and mix it with the fish in the bowl to make a good mixture.

Beat the eggs and mix it with the milk.

Put the fish/rice mixture in a heat-resistant dish in which you have spread the butter. Pour the milk/egg mixture over it and put it in the oven for 40 minutes.

The transformation of salt continues. In a sure sign that salt is getting posh, master chefs no longer talk of plain salt, they talk of its provenance.

On France's Atlantic coast, there is a shallow bay called Guerande. The salt from here, collected in the usual salt pans, is called the Flower of Guerande. The crystals here are skimmed off only in the evenings of really hot days, only when the breeze is deemed right and the work is done only by women. It seems as good a marketing gimmick as any, but aficionados claim that the salt smells faintly of violets and that it is more subtle, less salty and less aggressive than ordinary table salt which 'burns the food it touches'. There is, as expected, a hefty mark-up in price, around forty times in fact of table salt.

There is an even more expensive variety called *Fleur de Sel* (Flower of the Ocean) which costs over a hundred times the price of ordinary table salt. Apparently, it has a distinct and delicate flavour and mineral characteristics which accentuate the taste of seafood or vegetables. It is used for finishing only; regular sea salt is used for cooking.

Salt, as we have seen, has been many things to many people at many times in many countries. Unexpectedly, it has also been used as an argument to justify male chauvinism. This passage was printed as a news report in England before the First World War when suffragettes were fighting to get the vote.

> It has been left to scientists to establish that man is literally the "salt of the earth". Two famous French savants have just announced the result of a long series of investigations, which convinces them beyond all question of doubt that woman is unalterably man's inferior, because of the smaller percentage of chloride of sodium in her blood.
>
> In other words, the blood of the male has more salt than that of the female, and observations of animal life show that the more salt there is in the blood the higher the intelligence and general development. The indictment does not end there, for these savants declare that their combined physiological and psychological investigations have proved that woman is inferior to man in everything -- intelligence, reason, and physical force. The facial angle of the female, they add, more closely resembles that of the higher animals than the male, while women's senses are less keen than those of man and she feels pain less.
>
> The scientific explanation is that the blood of the female is poorer in red blood corpuscles, and therefore relatively poorer in brine, which has been found to be the important factor in the development of the individual.

Ernest Jones, a British psychoanalyst, quotes the passage in a paper on salt. He uses it as one more example to buttress his central thesis, which is that the massive importance given to salt through history in completely differing cultures is due to a very basic reason: salt is a symbol for semen. He says:

Salt is a pure, white, immaculate and incorruptible substance, apparently irreducible into any further constituent elements, and indispensable to living beings. It has correspondingly been regarded as ... the quintessence of life, and the very soul of the body. It has been invested with the highest general significance ... was the equivalent of money and other forms of wealth ... In religion it was one of the most sacred objects, and to it were ascribed all manner of magical powers ... To be without salt is to be insipid. The durability of salt, and its immunity against decay, made it an emblem of immortality. It was believed to have an important influence in favouring fertility and fecundity, and in preventing barrenness. The permanence of salt helped to create the idea that for one person to partake of the salt of another formed a bond of lasting friendship and loyalty between the two, and the substance played an important part in the rites of hospitality. This conception of a bond was also related to the capacity salt has for combining intimately with a second substance and imparting to this its peculiar properties; including the power to preserve against decay; for one important substance – namely, water – it had in fact a natural and curious affinity.

... If the word salt had not been mentioned in the preceding description anyone accustomed to hidden symbolism, and without this experience, would regard it as a circumlocutory and rather grandiloquent account of a still more familiar idea – that of human semen.

Jones has given many examples of the recurrence of salt as a motif in sexual rituals across the world. They are fascinating, whether we accept or reject his theory of salt being a symbol for semen. The theory certainly seems shaky when we consider the many superstitions and myths regarding salt as a good luck charm or salt as a protection against the dark forces. (Why on earth would anyone throw semen over the left shoulder at a witch?)

The fecundity of the sea – in giving limitless food, in the huge number of offspring of most sea creatures – and its saltiness gave rise to the belief that somehow it was the salt which helped in fertility. Sailors, for example, had observed that on ships carrying salt, mice bred in profusion. They, therefore, began to believe that female mice could beget young without the help of male mice.

Other superstitions have to do with marriage and the wedding night. In the Pyreness, a couple will put salt in their left pockets before setting off to church for their wedding. In some other countries the bridegroom alone does this, in others, the bride alone keeps the salt. In Germany salt is sprinkled into the bride's shoe while in Scotland, the salt goes on the floor of the nuptial bedroom.

The strong linkage between salt and sex also comes into language. The Romans called a man in love 'salax', from the word for salt. ('Salacious' has descended from salax). Shakespeare uses it in a similar way in *The Merry Wives of Windsor,* 'Though we are justices... we have some salt of our youth in us.' In Gujarati they say, *'Aa manas molla chhe. Ene ma koi meethu nathi'* (This man is insipid. He has no salt in him). Here, insipidity implies impotence. The opposite comes across in the Hindi phrase, *'Woh ladki namkeen hai'* (That girl is salty, i.e., sharp). In France, a menstruating woman was said to be *en salaison*, curing in salt. (She wasn't allowed to be present during the salting of food.)

In some parts of India, a woman wanting a male child fasts on the fourth lunar day of every dark fortnight and breaks her fast only on seeing the moon. A dish of twenty-one balls of rice, only one of which contains salt, is placed before her. If she picks up the rice-ball containing salt, she will have a son. If she doesn't pick up the salted ball, she can try again on the next suitable day.

The sex-salt linkage also comes through the practice of abstinence. Egyptian priests, who were required to be celibate at all times, were also required to abstain from salt for specified periods. Amongst tribes of Native Americans, the Dyaks returning from a head-hunting expedition or Pimas who had killed an Apache, were required to avoid both sexual relations and consumption of salt for several days.

James Frazer in *The Golden Bough*, originally published in 1890, gives the following examples of salt-abstinence from India: 'When a Hindoo maiden reaches maturity she is kept in a dark room for four days and is forbidden to see the sun. She is regarded as unclean; no one is allowed to touch her. Her diet is restricted to boiled rice, milk, sugar, curd and tamarind, without salt.

'Similarly, after a Brahmin boy has been invested with the sacred thread, he is for three days strictly forbidden to see the sun. He may not eat salt...'

Frazer also writes about the 'divine kings of Latin America': The heir to the throne of Bogota in Columbia, South America, had to undergo a severe training from the age of sixteen; he lived in complete retirement in a temple, where he might not see the sun nor eat salt nor converse with a woman.

We can conclude by agreeing with Jones' argument. Or we can disagree by looking at the straightforward explanation that salt was important per se and to the primitive mind, its diverse and multiple uses, its then lack of accessibility, its indispensability in food, the wars fought over it and its association with fertility and fecundity, imbued it with near-magical powers. Salt then lent itself to myth-making and superstitions, which resulted in its use in religious ceremonies. That, in turn, established it even further as a magical substance. We can now be more objective. But we still can't do without it.

LITERATURE, ART AND THE CELLARS MARKET

Why dost thou shun the salt? That sacred pledge,
Which, once partaken, blunts the sabre's edge
Makes even contending tribes in peace unite,
And hated hosts seem brethren to the sight!

Byron, *The Corsair*

Salt's importance through history is so well-established that one would expect references to it in literature. This is especially so because in alchemy, salt was looked on as one of the three basic elements out of which the seven noble metals emerged, giving salt a near-mystical status: mercury symbolized the spirit, sulphur the soul and salt, the body. In addition, mercury represented the act of illumination; sulphur, the act of union and salt, the act of purification. No wonder Herrick, in *Hesperides* said:

> The body's salt the soule is, which when gone,
> The flesh soone sucks in putrefacttion.

Other writers and poets too have used salt as a symbol and a metaphor. Like Charles Dickens who, uncharacteristically, wrote a Victorian ghost story in 1865 called 'To Be Taken with a Grain of Salt,' or George Gissing, who wrote a touching story about a man so good, he was always used by his friends, 'The Salt of the Earth' (1906). Some have written about it directly, none better than Adam Roberts in his SF novel *Salt* about a Noah's Ark-like escape into space:

> Salt is crystal compound of sodium and chlorine; faceted and transparent. Simple and pure. What life could there be without salt? It is known as God's diamond, by which we should be aware of the infinite variability of scale for the divine perspective. This tiny fragment of halite, it is a dot, an atom; but to God it can never be lost, it can never be overlooked or unnumbered. Every grain is a landscape, a world. It is a great cliff, a diamond as big as a mountain, a massive cube of ice. In it are embedded woolly mammoths, grimacing men in hides and skins. Buildings, cars, smooth as polished plastic, plain as glass.
>
> And salt combines the good and the evil, yin and yang; God and the devil. Take sodium, which is the savour of life. Without corporeal sodium the body could not hold water in its tissues. Lack of sodium will lead to death. Our blood is a soup of sodium. And here is the metal, so soft you can deform it between your fingers like wax; it is white and pearl, like the moon on a pure night. Throw it in water, and it feeds greedily upon the waves; it gobbles the oxygen, and liberates the hydrogen with such force that the hydrogen will flame up and burn. Sodium is what stars are made of. Sodium is the metal, curved into rococo forms, that caps the headpiece and arms of God's own throne. But here is chlorine, green and gaseous and noxious as hell's own fumes. It bleaches, burns, chokes, kills. It is heavier than air and sinks, bulging downwards towards the hell it came from. And here are we, you and me, poised between heaven and hell. We are salty.

Salt has also inspired poetry, like this wonderful poem by Pablo Neruda, superbly translated by M.S. Peden:

ODE TO SALT
(ODA LA SAL)

This salt
in the saltcellar
I once saw in the salt mines.
I know
you won't believe me,
but
it sings,
salt sings, the skin
of the salt mines
sings
with a mouth smothered
by the earth.
I shivered in those
solitudes
when I heard
the voice
of
the salt
in the desert.
Near Antofagasta
the nitrous
pama
resounds:
a
broken
voice,
a mournful song.

In its caves
the salt, moans, mountain
of buried light,

translucent cathedral,
crystal of the sea, oblivion
of the waves.
And then on every table
in the world,
salt,
we see your piquant
powder
sprinkling
vital light
upon
our food.
Preserver
of the ancient
holds of ships,
discoverer
on
the high seas,
earliest
sailor
of the unknown, shifting
byways of the foam.
Dust of the sea, in you
the tongue receives a kiss
from ocean night:
taste imparts to every seasoned
dish your ocean essence;
the smallest,
miniature
wave from the saltcellar
reveals to us more than domestic whiteness;
in it, we taste infinitude.

One of Aesop's fables features salt as a main ingredient.

THE SALT MERCHANT AND HIS ASS

A pedlar drove his ass to the seashore to buy salt. His road home lay across a stream into which his ass, making a false step, fell by accident and rose up again with his load considerably lighter, as the water melted the sack. The pedlar retraced his steps and refilled his panniers with a larger quantity of salt than before. When he came again to the stream, the ass fell down on purpose in the same spot, and, regaining his feet with the weight of his load much diminished, brayed triumphantly as if he had obtained what he desired. The pedlar saw through his trick and drove him for the third time to the coast, where he bought a cargo of sponges instead of salt. The ass, again playing the fool, fell down on purpose when he reached the stream, but the sponges became swollen with water, greatly increasing his load. And thus his trick recoiled on him, for he now carried on his back a double burden.

– Translated by George Townsend

There was a time when the use of salt in food was said to make you either irritable or melancholic or both. John Lyle, who described himself modestly as Maister of Arte, put forward this sentiment in *Euphues and His England* (1580):

> In sooth, gentleman, I seldome eate salte for feare of anger, and if you give me in token that I want wit, then will you make cholericke before I eate it; for women, be they never so foolish, would ever be thought wise.
>
> I staied not long for mine answer, but as well quickened by her former talke as desirous to cry quittance for her present tongue, said thus: If to eat store of salt, cause one to fret; and to have no salt, signifies lack of wit, then do you cause me to marvel, that eating no salt, you are so captious; and loving no salt, you are so wise, when indeed so much wit is sufficient for a woman, as when she is in the raine can warne her to come out of it.

There's even a clerihew about salt. A clerihew is a short, comic (or even nonsensical) verse about a famous person. It is written as two rhyming couplets with lines of unequal length, with one of the lines generally being the name of the person. It is named after Edmund Clerihew Bentley (1875-1956) who invented the form in the 1920s:

> Sir Humphry Davy
> Abominated Gravy.
> He lived in Odium
> Of having discovered sodium.

Folk tales about salt come from all over the world. Most of them illustrate the importance of salt. But this Chinese one tells you about how salt was discovered. The Chinese, after all, discovered everything. That was discovered by the Chinese too.

A PHILOSOPHICAL TALE

Everyone in China knows that the phoenix, or *feng-huang*, as it is known, is a beautiful bird, with its tail as bright as a peacock's and its scarlet head and breast and back. The *feng-huang's* wings are huge and colourful, and its eyes are as blue as the sea. The *feng-huang* is not only beautiful; it is also a noble and wise creature. It seldom appears, but everyone knows that when it does, it hovers over treasures, bringing fortune to those who see it.

One day a poor, hardworking peasant walked to his marshy fields for a long day's work. Suddenly he stopped and his eyes opened wide, for in front of him, half-hidden among the reeds, stood the fabulous *feng-huang*.

The peasant quickly ran toward the marsh, but as he reached the spot where the creature stood, it soared into the sky. The peasant watched it disappear, and then he turned to the spot where the *feng-huang* had been sitting. He smiled. 'There must be treasure buried here,' he said, and he began to dig as fast as he could.

He dug and dug, but he turned up only dirt and mud. At long last, he picked up a piece of earth and pondered. 'This dirt must be the treasure,' he said, and gazed up to the heavens. 'The *feng-huang* promises treasure,' he said softly. And so he wrapped the piece of earth in cloth and hurried home.

When he ran through the door, he called to his wife, 'I have found treasure,' and he sat down and told her his tale.

The two stared in wonder at the piece of earth.

'Dear husband,' his wife said after a while, 'you know you must take this to the emperor.'

The man nodded, for he knew, like everyone else in his country, that anyone who found a treasure must report it to the emperor. The peasant dressed in his work clothes, for these were the only clothes he owned. His wife carefully wrapped the piece of earth and placed it in a willow basket. Then the peasant took the basket in his hand and walked all the way to the capital city. There he announced his wish to present a treasure to the emperor.

When the emperor asked to see the gift, the peasant bowed low, reached into his basket and held out the earth. He told the emperor the tale of the magical phoenix.

The emperor frowned. 'You are trying to make a fool of me,' he cried. 'This is no treasure. Guards, take this man to the dungeon and put him to death. No one tries to trick the emperor!'

The emperor's guards obeyed their master. As for the basket of dirt, one of the servants placed it upon a shelf in the royal kitchen, and there everyone soon forgot all about it.

Some time later, one of the cooks was carrying a bowl of soup into the royal dining hall. As he walked, he passed beneath the basket, and a small clod of earth splashed into the soup. The cook was horrified, but just then the emperor boomed, 'Bring me my soup!'

The cook quickly carried the bowl to the table and placed it before the emperor. His hands trembled and sweat poured from his brow as the emperor dipped his spoon into the soup. The emperor took one taste and smiled. 'Delicious,' he said. 'This is the best soup I have ever tasted! What did you add to it?'

Still the cook trembled. 'Your majesty,' he began, 'I did nothing special, but a bit of dirt from the peasant's basket fell into the soup.' As he spoke, he turned as pale as the clouds.

The emperor was amazed. 'Bring me that basket,' he called to his servants, for he remembered the peasant's tale of the *feng-huang*. When the basket sat before him, the emperor reached in and sifted the earth through his hands. As he did, tiny white crystals clung to his palms.

'This is a treasure,' the emperor said. 'It is a gift from the phoenix. From this day on, we shall add these crystals to all of our dishes.'

He sent his men to dig in the earth where the peasant had first spied the phoenix. And that was how the people of China discovered salt and all its wonders.

The emperor wept for the peasant he had punished with death. He sent for the man's wife and son. He placed the peasant's son in charge of all the lands where the white crystal gleamed in the soil. The young man became rich and comfortable, and he cared well for his family.

And so the peasant, honoured through his son, rested in peace, and the *feng-huang* brought salt to China.

The Russian folk tale which follows is a variation on the King Lear-Three Daughters' story. The pearl tears are a particularly nice touch:

Once upon a time, there lived a woman on the top of a mountain in a cottage and had geese. In the large nearby forest, she would pick grass for the geese and fruit to carry home. One morning, a handsome young Count came into her presence. He asked if she had no one to help her carry her things. She told him that she was poor and had no one to help her and asked if he would be so kind since he was strong and tall. He agreed but soon he was groaning under the weight. 'These are so heavy, can we rest,' he asked? 'No. Go on a bit more,' she coaxed. He tried and tried to take the bundle from his back and found he could not. He began to think she was a witch. As if she could read his thoughts, she tried to console him by saying: 'Don't get angry. I will give you a present when we get to my home.'

Soon they arrived at her little cottage. It was a bit rundown though neat and tidy. There was another woman there who asked: 'Kind mother, you have stayed away for so long. You were missed.' 'I met this kind gentleman, who carried my burden,' replied the old woman as she took the bundles from the Count. 'Sir, you may rest upon the bench. And you, little one, go inside the house lest he fall in love with you.' The Count was somewhat surprised at the old woman's comment. The girl was homely and old looking and he thought love was an impossibility unless she was considerably younger.

The Count fell asleep. When he awakened the old woman was there ready to give him his reward for his kindness of carrying her bundles. She placed an emerald green box in his hand and admonished him to take good care of it. He put the unopened box into his pocket and left. He was unable to find his way out of the forest even though he had been able to before. Finally, after three days, he came to a large town. He was greeted by a guard who was instructed to take all strangers to the king and queen.

He respectfully explained his situation: 'Your Majesties, I am a Count. I have lost my way.' The king asked, 'How can you prove what you say?' The Count began to search his pockets and found the emerald box and presented it to the queen. Upon opening it she gasped in surprise and fainted. The guards seized the Count and the king helped the queen. As the Count was being taken away she awoke and asked that he be released for she wished to talk to him ... alone.

Once alone the queen began her sad tale. 'I have three daughters. The youngest was rare and wonderful. When she cried, pearls fell from her eyes instead of tears. One day, their father, the king, decided to divide his kingdom so he called our daughters before us. He said: "All of you love me. But she who loves me best will receive the greatest part of the kingdom."

The queen continued, 'Each child giggled and said that she loved her father best but was asked to tell how much. The first daughter said she loved her father as much as the sweetest sugar. The second daughter said that she loved him as much as her prettiest dress. Our youngest was quiet. The king asked her, "How much do you love me?" She replied, "I know not what to compare my love to, Father." He encouraged her and asked to think again. "I do not like even the best food without salt. Therefore, I love my father like salt." He became angry not understanding the compliment she had given him for salt is worth more than gold sometimes.

'"Like common salt," he raged! He had the kingdom divided between the two oldest daughters and placed a sack of salt upon her back and she was led into the forest by two guards. I begged him not to but he wouldn't change his mind. I wept. She also wept and the road to the forest was strewn with the pearls from her eyes. After a few days, the king regretted his behaviour and the soldiers were sent into the woods to find her. They could not. We have wept since.'

And so ended the queen's sad tale. 'When I opened the emerald box, I saw the pearl that my daughter used to cry. Where did you get it? the queen implored. 'In the forest I met an old woman and carried some bundles to her home. I didn't see a beautiful princess.' When the king was told of this, the three of them returned to the forest to look for the old woman. She was in her cottage spinning with the homely child beside her. An owl came to the window and the old woman said, 'It's time to go the well.' Off she went deeper and deeper into the forest. She brought up a bucket of well water and began to wash her face. As she did so, the homely mask soon came off and in the moonlight you could see she was the beautiful princess.

Meanwhile, the Count had strayed from the king and queen and climbed a tree to find them. But what he did see was the girl, a beautiful girl. He edged out further on the branch to secure a better look but the tree limb creaked. The girl heard the noise and placed on her mask as she ran from the well. He recognized her as the goose girl from the old woman's cottage. He climbed down the tree as quickly as possible but the fair maiden had disappeared.

He found the king and queen and said, 'I think I have just seen your daughter. She probably went down this path.' The three went hurriedly down the path and came upon the old woman's house. They peered in the window and saw the old woman alone at her spinning wheel. They knocked softly and heard her response: 'Enter. I was expecting you.' They asked the old woman if she knew of her daughter, the princess. The old woman rose from her stool and pointed a finger to the king and said, 'Three years ago, you unjustly drove her away. She who was good, kind and pure as salt!' She put out her hand, which was filled with salt and asked, 'Do you know the value of salt and therefore the love your child has for you?' The king expressed his sorrow and beseeched the old woman to show him his daughter. A door opened and the princess appeared. Everyone wept tears of joy but only the princess wept pearls.

The king asked her forgiveness and said that he had no kingdom left to divide and that he had nothing of worth to give to her. The old woman said: 'This child needs nothing. She is as the salt of the earth, pure, life-giving and watched over. Her pearls are finer than those of the sea and she shall always have them.' Upon this comment, the old woman put up her hands and said that for the years the princess spent tending her geese, the cottage was hers to keep. The kindly woman disappeared and the cottage changed into a beautiful palace.

In all of the commotion, the Count was overlooked and he began to go. The queen stopped him and asked if there was any way that they could repay him for finding their daughter. The king offered his gold, the queen offered the pearls. He looked at the princess and asked if she would marry him. The princess agreed...

And they all lived happily ever after.

And, finally, a feminist poem using salt, a fitting answer to the 'Men-are-superior' and the 'real' salt of the earth diatribe mentioned in the previous chapter.

A Woman of Salt
by
Charlene Elizabeth Fairchild

PART ONE

> I am a woman of salt
> My name is Eve.
> I am woman, risk-taker, rule-breaker, full of life, full of salt.
> Exulting when my first son was born I exclaimed:
> 'I have produced a man with the help of the Lord.'
> I am 'mother of all living' and acquainted with grief.
> I am salt of the earth.
> I am a woman of salt.
> My name is Sarah – the woman who laughs:
> 'Shall I indeed have pleasure and bear a child now that I am old?'
> Upon me rests the blessing of salt:
> 'She shall give rise to nations, kings and queens of peoples shall come
>   from her.'
> I am the woman of salty tongue.
> 'Get that other woman gone.'
> Sometimes my salt blisters.
> I am a woman of salt.

My name is Hagar, my tears are full of salt.
'God, my God, is a God-Who-Sees'
through my tears, through my fears, through my trials.
I lift up my voice and weep salty tears.
'What troubles you Hagar? Do not be afraid...
Come, lift up the lad and hold him fast with your hand
And I will make of him a great nation.'
God opened my eyes
And now I see a well of living water through the salt of my tears.
I am a woman of salt.
My name is Leah - the unloved.
Yet,
'The Lord has looked on my affliction' and given me a son, Reuben.
'The Lord has heard that I am hated' and I have a son, Simeon.
'Now my husband will be joined to me' for I have a son, Levi.
'This time I will praise the Lord' for my son Judah.
'Good fortune!' again a son Gad.
'Happy am I!' with a son, Asher.
' God has given me my hire' and a son, Issachar.
'God has endowed me with a good dowry' and my son, Zebulun.
And, now finally, a daughter, Dinah.
The salt of the waters of life has healed the hurt in my heart.
Now I am honoured in the building up of the house of Israel, the house of the Lord.
I am a woman of salt.
My name is Lot's wife.
You may remember me.
I am a pillar of the house of Israel, a pillar of the church.
I stand looking back, wistful and sad.
I am a woman locked up within my burning fears and drowned hopes.
My salt is the salt of bitterness.
But it is with bitter herbs and unleavened bread that Passover is remembered.
Remember me.

I am the salt of memory.
We are women of salt.
Women of the covenant,
Women of blessing.
Help us, Lord, to be salt in this world.
Salt full of flavour, full of healing.
Salt fit to use.

PART TWO

I am a woman of salt and gritty sand.
I am a woman of leaven, of pregnant yeast.
I am a woman of spice, of mystery and myrrh, of aloes and ointments.
I am a woman of blood, of life.
Ancient memories course through my veins.
I am a woman of light and shadows,
Silhouettes and starbursts, sunny smiles and fleeting frowns.
I am a woman of dark.
Night calls to me yet repels my very soul.
I am a woman of stones, of mountains and majesty,
of precious jewels and ancient fossils.
I see beauty in a lump of coal and stories of the universe in granite.
I am a woman of passion, of lust and love, of agony and hate.
He who made me, made me whole.
I am a woman of warp and woof,
Bright strands of colour and unbleached wool.
I weave memories both light and dark.
I capture moments that touch the heart.
I am a woman of word and song.
I am a woman of memory.
I capture stories and paint emotions.
Music, wild and full of fury, speaks deep inside,
fans the faint embers, stirs the ashes,
makes the flame burn bright.
I am a woman of salt.

Who would imagine that the substance which finally ends up in our salt shaker, could also at some stage in its life be used for high art?

Many of the once-busy and productive salt mines have now opened their doors – if mines have doors – to a different kind of world.

Gone are the hard-working miners, in some of them either working as slave labour or as prisoners or as captured enemy soldiers. The people who now walk in have gold credit cards in their pockets and cameras around their necks.

In the UK's Cheshire district, there's a salt museum as well as Lion Salt Works, both of which welcome visitors. Salzburg in Austria, has all its four salt mines now turned into tourist attractions. The journey into the salt mines begins with tourists wearing white overalls. Entry into the salt mine tunnels is on a small train. Tourists then go down smoothly polished slides from one level to the next till they reach the heart of the mountain. A raft then takes them across the famous salt lake.

Berchtesgaden was once Hitler's retreat. The nearby mines have now become a tourist spot, drawing 400,000 visitors every year. Entry is by a small train through a six hundred metre tunnel which takes you to the Kaiser Franz Sinkwerk, an enormous hall of three thousand square metres. Slide down a thirty metre chute to a beautiful salt cave. Here the near-transparent rock salt shows up in different colours in the lights. There's a salt lake here too on which a raft will take you to a sparkling spring.

Bolivia's salt producing region is a tourist attraction with salt-bearing caravans of llamas. This region also has a hotel made entirely of salt! It was built in 1993, has fifteen guest rooms, dining and living rooms and a bar. The hotel's walls are made of salt blocks stuck together by a salt 'cement' (salt mixed with water). In the dining room, not only is salt always on the table, salt *is* the table. So is most of the furniture. It must be the only place in the world where the management request the guests not to lick the walls or the furniture!

Other countries have also turned their old salt works into tourist spots. In the Chinese region of Sichuan, there's the Zigong Salt Industry Historical Museum. In Columbia, there's a cathedral carved into the rock salt. And in many of the mine shafts, tourists will find sculpted chapels and stations of the cross.

The most incredible of these transformed salt mines are in southern Poland. Situated at Wieliczha, near Krakow, these salt mines enriched a succession of Polish monarchs for six centuries before refrigeration and cold storages reduced the market for their produce. For the last hundred years, its operations have been winding down; simultaneously, it has been opened to tourists who now come from all over the world. And why wouldn't they, because the Wieliczha mines are an underground treasure house quite unlike anything else in the world.

All this started out, probably, as an early form of enlightened HRD. In the sixteenth and seventeenth centuries, the human cost of mining salt was extremely high. Records show that as many as ten per cent of the workforce was decimated every year. Some miners died of sheer exhaustion, because of long hours under difficult conditions. Others were killed in accidents from methane explosions, or collapsed shafts. Hundreds of horses used to move the salt died too. Not surprisingly, each shift began with prayers for safety. In 1689, the mine employed a priest to hold a daily Catholic service in an excavated salt chamber. This came to be called St Anthony's Chapel after the patron saint of miners. The miners were encouraged to decorate this space so that it would actually look like a chapel.

From this small beginning grew the wonders of Wieliczha, as miners with artistic inclinations, and later professional artists from outside, began to make free-standing statues and carved bas-reliefs with religious motifs wherever they could. Stunningly, the material used for all these was salt.

They went beyond that. Whole chapels were carved out of rock salt, like The Princess Kinga Chapel. Not only was it carved out of salt, but also so was everything in it : the altar, floor tiles, bas-relief on the walls... Even the chandeliers are made from salt rock crystals. The salt statues inside it and at its entrance have been described as majestic. Elsewhere, there is a delicately carved bas-relief of 'The Last Supper' made of salt and a salt statue of Pope John Paul II. These statues might look like rock statues till you get a light behind them; then you see their translucence. To complete the picture, even the stairs and furniture is made of salt.

There's more artwork inside. And inside. After all, the mines contain 7.5 million square metres of excavated space on nine levels, going down from sixty-five metres to a final depth of 335 metres (1100 feet). The underground passages add up to a staggering 320 km, while the number of chambers total 2148. The tallest of these is thirty-six metres (118 feet). Two others are thirty metres (ninety-eight feet) high. Kinga itself, located 101 metres (331 feet) below the surface, is fifty metres long, fifteen metres wide and twelve metres high (164 x 49 x 39 feet).

The church has been used for symphony concerts and even state banquets and balls. The highest chamber, the 'Warszawa Chamber' has been used for a Guinness record bungee jump, as a theatre and as an opera hall. The complex also has three underground salt lakes, the deepest reaching seven metres (23 ft). One of them has been equipped with a wave machine.

Salt, however, has an ambiguous relationship with water. Although they are usually complementary to each other, in the context of Wieliczha, water in the form of moisture has become salt's adversary, damaging a lot of the art works and reducing many of its statues to ruins. Those near the entrance, like the St Anthony's Chapel, have been particularly affected.

UNESCO recognized Wieliczha as a World Heritage site, and in 1978 put it on its first endangered list. Since then, much money has been spent in micro-climate control and the danger from excessive moisture caused by humidity and the body heat of visitors, has been brought under control. Wieliczha then became the first site to be taken off the endangered list.

Gertrude Stein might have said 'A cellar is a cellar is a cellar.' But the once humble receptacle for salt soon became an ornate thing. It became an heirloom, a craftsman's showcase, and finally, a social arbiter which told a guest at dinner exactly where he or she belonged.

This was clear from the seating arrangements – at banquets (described in an earlier part of the book). At the head of the table, sat the host. In the centre of the table, some half way down its length sat the salt cellar, or as it was called, The Salt. If you were placed 'above the salt', i.e., between the cellar and the host, your social status was assured, and openly acknowledged. 'Below the salt' you really weren't important. It is said that during the reigns of Henry VII and Henry VIII, ushers were specially trained in the difficult task of seating people by their rank. What happened to the usher who bungled? Did Henry VIII chop his head off? Not really, the guest was moved to his appropriate place, which must have been pretty humiliating. But in earlier eras, the eleventh century for example, a law laid down by King Canute decreed that any guest who presumed to sit above his station could be 'pelted out of his place by bones, at the discretion of the company, without the privilege of taking offence.'

By the fifteenth century, the laying of the table itself had become an elaborate ceremony in royal households. First was placed the 'Chiefe Napkin'. Then officials in charge of such matters brought the salt and they 'set the salt right under the middest of the cloth of estate.' Queen Elizabeth the First had a ceremony which became almost like a Dance of the Salt: after the tablecloth was laid, two men arrived, one with a rod, the other with a cellar. They knelt, then placed the rod and cellar on the table. They were followed by a 'virgin' dressed in white and a maidservant with a knife for the salt. The virgin then took the knife and placed a portion of salt on each plate.

There were, obviously, some tricky questions to resolve when you had many guests of importance. The story goes that Charles V of France entertained Charles IV of Rome and King Wenceslaus of Germany at one meal. Scope for a bit of diplomatic outrage here, possibly leading to conflict, even war. The solution: an elaborate salt cellar was placed before each king. By the late seventeenth century, elegant households had found an elegant answer: the large and elaborate cellars had become small and elaborate salt dishes, individually placed, one before each guest. No one was below-the-salt then.

Royal and state cellars, though, remained rather large and rather ornate. One dating back to Wales in 1586, for example, was an eighteen inches high (0.45 metre) cylinder on top of which was a vase, and which was decorated with sterling silver fruit, birds and animals. There were six-sided cellars depicting the Labours of Hercules. There were large swans, hens in nests, dogs and even an 'olifaunt'. There were lavishly decorated thrones with the seat being the lid from the court of Tsar Nicholas II. In the Tower of London, in the room containing the crown jewels, are eleven magnificent gold cellars, some from the reign of Elizabeth I, the most elaborate, in the shape of the tower, from the reign of Charles II. The latter was used only at coronation banquets, which means very sparingly. Many state cellars were in the form of a ship. The Ship of State. The Salt of State...

There were many with religious motifs, though these were generally in private (and rich homes). They might be cast in pewter with scenes of the Annunciation and the Latin inscription ' Hail Mary full of grace, the Lord is with thee'. There are others featuring Abraham and Isaac or Cain and Able. Some have been found with the inscription 'When you are at table think first of the poor', which was odd, given the expense of the cellar.

People now collect cellars, called 'Open Salts' in America. You can get them on auction sites or on e-bay. You can join Open Salts clubs or hunt for them in antique shops. Faberge made them in gold, others in porcelain, crystal, glass, wood, pewter and silver. People now spend a fortune on them and keep them for the prestige (and price appreciation) they will bring them.

They went out of use because a number of things happened. Society changed and people became more egalitarian. Salt changed too: salt cellars were not shakers, they were containers from which you took salt with your thumb and index finger, or on the flat of your knife because salt wasn't the fine, running variety we have now, but pieces of rocky crystals. Salt changed in another way: it was no longer a precious, difficult-to-get rich man's status symbol. Instead it became an item of everyday use. It was absurd, then, to present it so lavishly.

In its purest form, without the hierarchical snobbishness which soon overtook it, the ceremony of salt was a rather gracious one. Salt was placed first on the table, before guests arrived, and removed last, only when guests had left. The guests were thus told, that they were 'invited in love and friendship' and were 'loved before they came'. Not removing the cellar signified that while feasting and eating may come to an end, love and friendship remained.

THE (LITTLE) BOOK OF SALT QUOTATIONS

Nemini fidas, nisi cum quo prius multos modios salis absumpseris (Trust no one unless you have eaten much salt with him).

Marcus Tullius Cicero, De Amicitia

Salt is born of the purest of parents: the sun and the sea.

Pythagoras

Salt is white and pure – there is something holy in salt.

Nathaniel Hawthorne

Let your speech be always with grace, seasoned with salt.

Colossians 4:6

Let yourself be open and life will be easier. A spoon of salt in a glass of water makes the water undrinkable. A spoon of salt in a lake is almost unnoticed.

Gautam Buddha

You shall find out how salt is the taste of another man's bread, and how hard is the way up and down another man's stairs.

Dante Alighieri

The precious salt, that gold of cookery!
For when its particles the palate thrill'd
The source of seasonings, charm of cookery! came.

Hesiod

Ye are the salt of the earth: but if the salt have lost his savour, wherewith shall it be salted.

The Testament of St Mathew 5:13

Timon hath made his everlasting mansion
Upon the beached verge of the salt flood;
Who once a day with his embossed froth
The turbulent surge shall cover.

Shakespeare, *Timon of Athens*

With all thine offerings thou shalt offer salt.

Moses, *Leviticus*

Salt is what makes things taste bad when it isn't in them.

Anonymous

Wit is the salt of conversation, not the food.

William Hazlitt

It takes four men to dress a salad: a wise man for the salt, a madman for the pepper, a miser for the vinegar, and a spendthrift for the oil.

Anonymous

Many are the ways and many the recipes for dressing hares; but this is the best of all, to place before a hungry set of guests a slice of roasted meat fresh from the spit, hot, season'd only with plain, simple salt ... All other ways are quite superfluous, such as when cooks pour a lot of sticky, clammy sauce upon it.

Archestratus

An honest laborious countryman, with good bread, salt and a little parsley, will make a contented meal with a roasted onion.

John Evelyn

Ham: 40 days in salt, 40 days hanging, in 40 days eaten.

Joseph Delteil, *La Cuisine Paleolithique*

Of all smells, bread; of all tastes, salt.

George Herbert

It is a true saying that a man must eat a peck of salt with his friend before he knows him.

Miguel de Cervantes, *Don Quixote*

Salt is the only rock directly consumed by man. It corrodes but preserves, desiccates but is wrested from the water. It has fascinated man for thousands of years not only as a generator of poetic and of mythic meaning. The contradictions it embodies only intensify its power and its links with experience of the sacred.

Margaret Visser

To stand at the edge of the sea, to sense the ebb and flow of the tides, to feel breath of a mist moving over a great salt marsh, to watch the flight of shore birds that have swept up and down the surf lines of the continents for untold thousands of years, to see the running of the old eels and the young shad to the sea, is to have knowledge of things that are as nearly eternal as any earthy life can be.

Rachel Carson

No man is worth his salt who is not ready at all times to risk his body–to risk his well-being – to risk his life – in a great cause.

Theodore Roosevelt

I really don't know why it is that all of us are so committed to the sea – except I think it is because in addition to the fact that the sea changes and the light changes, and ships change, it is because we all came from the sea. And it is an interesting biological fact that all of us have, in our veins the exact same percentage of salt in our blood that exists in the ocean, and therefore, we have salt in our blood, in our sweat, in our tears. We are tied to the ocean. And when we go back to the sea, whether it is to sail or to watch, we are going back from whence we came.

John F. Kennedy

Being kissed by a man who didn't wax his moustache was like eating an egg without salt.

Rudyard Kipling, *The Gadsbys*

Salt is the policeman of taste: it keeps the various flavours of a dish in order and restrains the stronger from tyrannizing over the weaker.

Margaret Visser

When life hands you lemons, break out the tequila and salt

Anonymous

A wise woman puts a grain of sugar into everything she says to a man, and takes a grain of salt with everything he says to her.
>Helen Rowland

Those who by unselfish lives and consideration for others elevate the tone of the community in which they live and who by their presence make others happier, these are the salt of the earth.
>William Lyon Phelps

Where would we be without salt?
>James Beard

SEE HOW IT RUNS

The story of salt's transformation from a rare, fought-over commodity to one of common, everyday use is a recent one. The story of salt in India, the transformation of an import-dependent country to a happily exporting one, is a story of two men of the twentieth century.

The way salt is made today is the way it has been made through history. That has to do with its source: the source is brine (salty water) from seas, salt lakes, salt springs and similar bodies of water. Even salt deposits that are deep under the ground were formed by evaporation of sea water millions of years ago. Evaporating sea water through salt pans is the oldest method of obtaining salt. It is the preferred method even today in countries like India which have an abundant source of saltwater, large tracts of land for salt pans (evaporating ponds) and a hot, dry climate with plenty of sunshine to speed up evaporation.

Other countries which have only sporadic sunshine had to boil the brine to evaporate water as they did in Europe or to design covers for their ponds in case of rain as they did in the United States. Inland countries with no access to the sea like Switzerland, Austria or Hungary, had underground salt deposits which could be drilled into.

Amazingly, not too much has changed. Ancient Egyptians got their salt from dried salt lakes and salt deposits. They invented a method of 'dragging and gathering' which has only changed in that it is now mechanized. The ancient Chinese method of drilling salt has only got a bit more sophisticated. The Romans used single ponds for solar evaporation; now, a series of ponds is used so that the brine which starts off at around 2.5 per cent salt from sea water goes from pond to pond of increasing salinity till it reaches super-saturation at twenty-six per cent. But even this method was adopted before the ninth century. What no one does now – not publicly at least – is use the old test to gauge concentration: place an egg in the salt pan. If it floats, the salinity is ok. Neither does anyone add, as they used to a few hundred years ago, a bit of blood from the butcher's and a bit of ale from the brewery to (allegedly) hasten evaporation. And, thankfully, no one uses forced labour to pump brine with pedal power as was done when conquering armies took slaves (China used cattle till 1902. In the Zigong salt works, there was a 'work force' of 100,000 oxen and subsequently plenty of beef). Even today movement of salt is by road transport which features camels (in India, West Asia) and llama (Bolivia), though mechanized animals in the form of trucks are gradually taking over almost everywhere.

However, once the raw material has been obtained by these unchanging methods, the production of salt gets highly sophisticated. For example, at its Mithapur plant, Tata Salt is produced through a process known as 'Vacuum Evaporation Process'. In this, saturated brine is taken to evaporators in which low pressure steam is used to evaporate the water from the brine. After achieving a specified concentration, the salt slurry is taken into decanters. From here, the solid slurry is taken to a centrifuge to remove the 'mother liquor' (the liquid portion). At the same time, a measured quantity of potassium iodate solution is added into the centrifuge along with the slurry so that iodization of salt is achieved uniformly and consistently.

During the centrifuging operation, a non-caking agent is also added to make the salt free flowing.

After the centrifuge, the salt is passed through a fluidized bed drier which makes the salt completely dry and free from moisture. It is then taken to stainless steel storage bins for packing.

The evaporator bodies are made from titanium to maintain a high level of hygiene. This evaporation process ensures high content of NaCl and also consistent quality and purity of salt. Edible salt produced through this process is generally considered much purer than refined salt.

Two simple changes made a large difference to the salt we eat today. In 1911, the US's leading producer, the Morton Salt Company, added magnesium carbonate to table salt. (Later, magnesium carbonate was replaced with another neutral, non-sticking chemical, calcium silicate.) This prevented crystals from sticking together, so that the salt ran freely. This led to Morton's famous ad campaign of 1914 showing a little girl standing in the rain with a large umbrella and a container of Morton's salt from which a stream of salt is seen spilling. 'When it rains it pours' was the ad's slogan. Some years later, an English company showed a little boy pouring salt over a hen. 'See how it runs' was its ambiguous slogan.

In 1924, Morton introduced another element into its salt. The company was advised by the state medical association to add a trace amount of iodine compound (sodium iodide, potassium iodide or potassium iodate) to make the world's first iodized salt. This prevented goitre, a swelling of the neck resulting from enlargement of the thyroid gland, which was then affecting a significant proportion of the American population.

Both procedures are now, more or less, accepted by all countries: the first, the use of a neutral non-stick agent is used routinely; the second has been a bit more controversial, some countries opting out because goitre wasn't one of their problems or because of the general prejudice against chemical additives. However, goitre being fairly widespread in India, iodization was made compulsory in 1998. But lobbying by small producers and Gandhians (who said *everyone* should have the right to produce salt) pressurized the government into revoking the ban on non-iodized salt in 2000. The position now is that each state can evolve its own policy. In the event, most states have opted for iodization, especially since it is a WHO recommendation. Tata Salt was the pioneer of iodization in India, producing it commercially in 1985, well before it was made mandatory. The process it uses mixes liquid potassium chloride in the wet slurry in the centrifuge itself. This gives a much greater consistency than the conventional method of adding the iodizing agent as a powder at the end of the process.

The journey to modernity has been an arduous one for the Indian salt industry. To start with, it had a most formidable adversary in the East India Company. When the British government took over from the company and formed an 'Indian' government, its adversarial role only got more sophisticated. But in Kapilram Vakil, it was up against a doughty fighter.

Raj Ratna Kapilram H.Vakil, BA, MSc Tech (Manchester), FIC, MI Chem E set up his own consultancy for the chemical industry on his return from England. In the early years (1913-16), he did some pioneering work for the Jamshed Oil Mills of the Tatas. That was followed by work at the Tata Iron and Steel plant in Jamshedpur. What he did was important, some of the processes he developed being path-breaking for industry internationally. But what he will be remembered for most was salt.

Through 1916-17, there was a real salt famine in Bengal and other eastern parts of India. As earlier chapters have shown, the East India Company had systematically killed the indigenous salt industry in Bengal and Orissa so that Liverpool and Cheshire salt could be exported there. Stringent 'border' control also ensured that no salt from other parts of India could enter these states either officially or unofficially. Bengal, Bihar and Orissa had thus become completely dependent on imported salt.

This had gone on for over a hundred years, but when the First World War started, there was a sudden, and acute, shortage of foreign shipping. Salt prices in Bengal, maintained for years at Rs 70 per 100 maunds shot up to unprecedented levels; first to Rs 300, then the high of Rs 434 per 100 maunds in 1917. Responding to the situation, the Tatas commissioned Vakil to carry out an exhaustive survey of salt sources on the East Coast, particularly on the coast of Orissa and along the coastline of Madras. The objective was to establish a salt industry large enough to do away with imports altogether.

There were a few false starts, with first Chilka Lake near Jaganath Puri, then Velan in Kathiawar (part of Baroda State at that time) being selected as possible sites. But an acute shortage of funds due to problems in the steel industry forced the Tatas to abandon both projects before they had started.

The breakthrough came in 1926, when the Maharajah of Baroda opened a port at Okha. He called Vakil to develop salt and alkali industries in Okhamandal. Vakil set to work, but progress was slow. The problem was again money: India, like the rest of the world, was going through a severe trade and industrial depression. Finally, the first shipment for Bengal was sent off on 5 May 1928. According to Vakil, 'For the first time Calcutta received fine white crushed salt of Indian manufacture comparable in quality with foreign imported salt.' The shipment was sold off completely within a few hours of the steamer arriving at Calcutta. The price: Rs 97 per 100 maunds, whereas the cartel of foreign salt importers had maintained prices in the Rs 100-105 range. The combine retaliated by dropping the rate to Rs 60 in a single day, which was later raised to Rs 92. A fierce rate-war had begun, the main antagonists being Okha and salt suppliers from Aden.

Salt manufacture in Aden, incidentally, had a strange history. The Aden Salt Works were founded in 1886 by the Burgarellas, two Italian brothers from Sicily. Things weren't easy at all as this contemporary statement shows:

> The brothers Burgarella were compelled to work under the worst conditions... and worked very hard personally, in a very dangerous climate, a hut was all the shelter for making the natives acquainted with the handling of simplest necessary tools.
>
> The founder Messrs Burgarella's father, was compelled to sell out many properties in Sicily, now worth millions of lire, to supply money for this industry, which was deemed a FOLLY, both in Aden and Sicily, and he died on his way home on account of the hard work, mishaps, anxiety and the climate of Aden.

Things got better later. Then Indian entrepreneurship, using Indian capital helped to expand Aden's salt industry. Aden was ideal for salt: there was no monsoon, so salt could be made through the year (compared to nine months in India); the temperature was high and uniform (compared to the low winter temperatures at Okha); high winds through the year helped the process of crystallization and enhanced the rate of evaporation; the initial density of the brine was as high as 8 degree baume compared to the normal 3.5 degree baume of sea water. And lastly, there were favourable duty concessions because Aden, in spite of its geographical location, was officially a part of British India!

By 1930, Aden's production figures were as follows:

Salt Works	Proprietors	Average	Production
Adam Salt Works	A.Burgarella	1000	125,000
Indo-Aden Works	Abdoolabhoy and Joomabhoy Lalljee	900	75,000

Hajeebhoy Salt Works	Hajeebhoy Lalljee	943	15,000
Little Aden Salt Works	Palonjee & Brothers	900	15,000
	TOTAL	**3,743**	**230,000**

Each of these had considerable scope for expansion, yet Baroda and Vakil took the decision to press ahead, reduce cost of production to the minimum per ton and simultaneously increase production to 35,000 tons.

Two interesting tables from that time give a clear picture of what Vakil achieved and why Sayajirao Gaekwad of Baroda conferred the Rajratna title, the state's highest, on him.

The first table is an:

ANALYSIS OF SALT MADE AT MITHAPUR

Standard analysis on dry basis

	Percent
Calcium Sulphate	0.93
Sodium Chloride	97.98
Magnesium Sulphate	0.21
Potassium Chloride	0.05
Magnesium Chloride	0.52
Insollubles	0.31
	100.00

The salt crystals (kurkutch) before they are sent to the mill are of about ¼ inch size and the ground salt has a fine grain of 1/64 inch.

(These figures confirm that the salt exceeded international standards.)

The second table tells us how successfully Okha reduced imports from the West into the country:

SALT IMPORTS INTO BENGAL

		(Per cent)
From	1924-25	1934-35
United Kingdom	17.38	...
Aden	33.13	51.78
Germany	5.57	6.68
Spain	4.71	...
Tunis (N. Africa)	0.91	...
Italian E. Africa	7.73	...
Egypt	27.13	0.38
French Somaliland	3.44	...
Karachi		15.09
Tuticorin		2.99
Ras Hafun		0.80
Bombay		3.08
Navlakhi		3.51
Porbander		4.42
Okha		11.27
	100.00	100.00
Quantity (tons)	543,597	514,443

The decision to press on with an increase in production, nevertheless, was a brave one. To start with, the competition, especially from Aden, was ruthless, ready to exploit any tax break to dump salt at seemingly suicidal prices. Even more than that was the transparently hostile attitude of the British-Indian government, which made sure there was no level playing field. First the government set up a commission on the salt industry.

The Strathie Report came out with the conclusion that 'the production of fine white salt could not be undertaken advantageously in India'. The second was a Central Board of Revenue report which told the government that 'it is impossible in the near future so to expand the output of salt in India as to render the country self-supporting.'

When the Maharajah of Baroda found he couldn't continue to fund this unequal battle at the level required, the end of the Okha Salt Works, and with it, the dream of an Indian salt industry, seemed near. In a last throw of the dice, Vakil asked JRD Tata to step in, but he went beyond salt and presented a blueprint for an integrated chemical plant, where salt would be one of the products. JRD flew in his own plane to look over the site. His dramatic entry from the sky pulled in villagers from miles who came to gawk at the 'Tata flying bird'. The first step had been taken; Tata Chemicals was registered in January 1939, with Vakil as its technical director. His objective: to manufacture caustic soda, soda ash, chlorine products, marine chemicals, potash and other products. By 1943 when he was sixty, Kapilram Vakil had transformed Okha (now named Mithapur), from a desert overrun by cactus, snakes and scorpions, into an industrial township supporting hundreds of families either through direct employment or indirectly. The foundation was laid for the infrastructure which made India a surplus salt country soon after Independence.

Three years later, Vakil died, with one dream still unfulfilled: in spite of his best efforts, he had been unable to set up a soda ash plant. The manufacturing process was a well-guarded secret of a handful of companies around the world and none of them was willing to part with the secret. Manufacturing involved fifteen different process and power units and successful operation demanded that each one of these units individually and together work right at the same time. Perhaps if Vakil had been around a few years more...

Tatas brought in foreign experts. The first was a Chinese

group led by Dr Te Pang Hou. He was followed by a world-renowned American consultant Zola Deutsche. For various technical reasons, he didn't think the project feasible. 'You are in the wrong place and in the wrong business. The sooner you get out the better,' was his general summing up. JRD's reply was that the Tatas never quit, 'not when we have aroused people's hopes.'

Serendipitously, that's when JRD ran into Darbari Seth, a 31-year-old chemical engineer who had just returned from the United States after a brilliant academic career and a highly successful stint working for Dow Chemicals. As it happened he had helped set up a soda ash plant in Holland. To cut a long story short, Seth worked on the design of the complex using a team of young engineers (average age twenty-six), all brilliant, all committed to the cause. 'He nearly demolished the place,' JRD then said, 'But things began to change for the better.' When Seth and his team were through, Deutsche was shown the blueprint. His reaction was simple. 'Anyone who can design like that, does not need to consult me or anyone else.' The soda ash plant was up and running; it soon exceeded its capacity of 400 tonnes. This was the foundation on which the Mithapur complex was built to emerge as the largest inorganic chemical complex in India, and one of the largest in Asia.

But when things are up, they have a habit of going down. In 1962, the monsoon failed completely: the Mithapur complex, relying for its freshwater supply on two lakes was facing a complete draught by October. Seth's reaction was combative. 'Mithapur will shut down,' he said, 'over my dead body.' He drove the company into emergency mode: every employee was roped in to come up with innovative ideas to cope with the situation. Some 250 ideas were put into action, ranging from conservation, to substitution to production, so that the plant not only survived, but flourished. In the following nine years, the monsoon failed seven times. But Tata Chemicals continued to grow. It was also able to supply drinking water to all of the forty-two villages in Okhamandal.

Today Tata Chemcials is the largest producer of vacuum evaporated salt and soda ash in India, with a market share of forty per cent and thirty-five per cent, respectively. Mithapur, in fact, is the largest salt works in one location in the whole of Asia, producing 500,000 tonnes per annum. When India exported salt to the USA, it must have made four men sitting up there – Sayajirao Gaekwad, JRD Tata, Kapilram Vakil and Darbari Seth – smile some very broad smiles. The country once deemed incapable of producing fine white salt, was now sending it across the seas to the world's biggest producer and consumer. You could call it a sweet triumph. Or more appropriately, a salty triumph.

LADIES OF SALT

The first Lady of Salt was, undoubtedly, Lot's wife. But we know what happened to her, yet all she did was give in to human curiosity. You would be made of stone if you didn't stop to look at God's show of fire and brimstone. She wasn't made of stone, so God made her into salt. More ignominy: she is now a tourist attraction near the Dead Sea. Worse, no one calls her by her name. She is always 'Lot's wife'!

Today's ladies of salt don't get punished for their curiosity. They are, instead, encouraged to stop and see, look and learn. They use their own names. Their literacy may only extend to learning their own signature (no more the humiliation of the thumbprint), but they can add and keep accounts and calculate accumulated interest. Their Gujarati now has a smattering of Hindi (a creeping pan-Indian identity) and this hybrid language is now spoken with a hitherto unknown assertiveness. Their boundaries are extending, even literally, as they travel for the first time from the confines of their protected villages to the big town, and even to the bigger metropolis. From time to time, they do look back: at how things were and how they have got better. And they are not turned to salt.

Mithapur is everything that a salt production centre should

be. It is on the coast, its water has high salinity, it's got low rainfall, non-stop sunshine and high wind velocities. It's also got a large expanse of open, flat land. But these characteristics make for a parched and dry land. In fact, it is one of the most drought-prone regions of India, and the living isn't easy. Yet, confirming once again that those who live on the land don't necessarily know how to live by the land, the region's forty-two villages had made traditional agriculture their main activity, cultivating crops like bajra and jowar that relied solely on rainwater. Villages also faced a severe drinking water shortage beginning every February and continuing until the onset of the monsoon. In most places, underground water is brackish beyond a depth of nine metres (thirty feet); in others, the land is rocky and the water table extremely deep. In short, water for drinking and irrigation is scarce.

Faced by a crisis, the concerned citizen's knee-jerk reaction is to raise funds for relief work, but relief work by its very nature is temporary. For sustained development, you need intervention from agencies which bring in funding and expertise and involve the people, so that projects become self-sustaining in the long term. Tata Chemicals had already been through this experience in the 1960s, when Darbari Seth led a successful programme for recycling and conservation of water and for substitution of rain water wherever possible by sea water.

The Tata Chemicals Society for Rural Development (TCSRD) was set up in 1980. Either on its own, or in collaboration with the state government's District Rural Development Agency (DRDA), TCSRD formulated a comprehensive water management and watershed development project. The objective was to conserve as much rain water as possible in as many areas as possible so that water would be locally available. This entailed building check dams, bunds, percolation tanks and farm ponds. Water-harvesting included the recharging of wells as well as the deepening and desilting of ponds and wells.

The difference this has made to villages and villagers is palpable. The land no longer seems hostile, bunds and small dams made of earth rise unobtrusively, creating catchment areas. A villager, unbidden, rushes off to bring a sample of large red and green chillies which is one of the crops now irrigated because of the dam. Groundnut, green gram among others, are also cultivated. Typically, one dam can support fifty farmers.

An increase in the standard of living is noticeable and the villagers' faith in these projects is clear from their willingness to contribute part of the cost, as well as pay for services received. Villagers are encouraged to form village committees, and these groups decide on the distribution, use and payment mechanism for water. Money raised thus has helped meet upto eighty per cent of the expenses of new dams.

The villages which have collectively gained from these projects, financially and otherwise, now have a collective clout when dealing with banks or district authorities. After all, the direct economic gain is of the order of Rs 2.5 crore per year. There are other direct benefits, like greater employment to landless labourers. The biggest plus, of course, is that it is the villagers themselves who are owners, custodians and beneficiaries of these schemes. There are also indirect benefits, which in rural areas constitute a minor social revolution, the kind Mahatma Gandhi was trying to launch throughout his life. This list gives the names of a village committee. The columns tell their own story: here's the beginning of a breakdown of the caste system, so often the bane of rural society.

No.	Name	Caste	Designation
1.	Khumbhabha Jethbha	Vagher	President
2.	Duda Devra Nagesh	Rabari	Vice President
3.	Somgar Mohangar	Bavaji	Secretary
4.	Ramji Mava Tavdiwala	Kharva	Member
5.	Rana Ala Vasra	Aahir	Member

6.	Chandubha Dipubha	Rajput	Member
7.	Latif Ishaq Chavda	Muslim	Member
8.	Kishor Jesang Rathor	Kharva	Member
9.	Hansaben Bhanjibhai	Kharva	Member
10.	Karman Sava Parmar	Harijan	Member
11.	Kara Juma Sora	Muslim	Member
12.	Abram Osman Bolim	Muslim	Member

There's yet another social revolution underway on the periphery of the industrial action in Mithapur. This involves the silent, often oppressed, neglected minority of rural India, its women.

The water projects of Okhamandal may have transformed lives, but they cannot change its monochromatic scenery which remains a dusty brown. Brightening it up, though, are Okhamandal's women in billowing skirts and blouses and *odhinis,* of bright earth colours, the brick red and indigo and terracotta and black that comes from natural dyes. These are striking-looking women, with a delicate bone structure and dazzling, often light-brown eyes. When they speak, they do so to an accompaniment of metal on metal, as silver bangles, chandelier earrings and necklaces move with their gestures. Then you notice the tattoos, which cover every inch of their hands and arms. Wasn't there pain and blood? 'Of course! Lots of pain and lots of blood. But what choice was there?' Later you notice that only some women have tattoos; it is a caste thing.

Tattooed or not, they were all in the same predicament as their men, before the water situation eased up. Worse, because they were the ones who had to walk miles in search of water, then walk miles back with water on their heads. And they were the ones who had to micro-manage whatever little water they could get. There's no running water yet, as an American visitor famously remarked, but things have got easier.

They got even better when TCSRD introduced the concept of the Self Help Group (SHG) into the villages of Okhamandal. The idea behind the SHG was simple, which is why it works: use the natural skills of local women; and channelise their natural instincts into a formal structure. The first part took the shape of handicraft development, the second was to introduce the idea of banking and micro-credits.

Okhamandal women already had the basic skills in handicrafts: they decorated their clothes with embroidery, they added beads and other embellishments, and they did all this as routine. The difference was that they did it for themselves. Through TCSRD they would do it for others, and earn a steady income from it.

A display outlet has been set up in Mithapur, which also serves as a distribution centre. Cut cloth is sent to the women in their villages and picked up as finished goods. Two hundred women in seventeen villages now work regularly on the programme; that number is expected to treble shortly. Some of these women, in fact, support unemployed husbands while others are widows bringing up children. Most supplement the family income. Sales have gone up to nearly four lakh rupees in 2003, twice that of the previous year. The range of products has expanded from the staple of bedcovers, tablecloths and cushion covers to now include shirts with appliqué work and short *kurtis*. There's even a brand name for the products: 'Okhai', meaning 'From Okhamandal'. Sales are through exhibitions at different centres and through women's associations of the armed forces.

There are over a hundred Self Help Groups, with a membership of nearly 1500 women. The majority of women may not yet be involved in handicrafts, but they are certainly immersed in their 'banking' venture: each woman saves a fixed sum of money every month. 'We started with Rs 10 per month. Now it is Rs 30 every month. We hope to get to Rs 50 soon.' The group banks this monthly sum, and then collectively takes a decision on the loan requests that come in: someone wants to buy a new sewing machine, someone else needs money for irrigation, or to plant potatoes or to set up a shop or a flour mill or to buy a buffalo. The loan limit is Rs 15,000 and each loan needs two guarantors. The repayment record has apparently been excellent.

The oldest group is ten years old. Its savings now exceed Rs 150,000. That figure worked as a kind of beacon: more and more women wanted to join in, and did, forming SHGs with a minimum of ten and a maximum of twenty-five women. Then something strange happened: the men decided that they wanted to start SHGs too. 'One lakh, fifty thousand?' they said, 'May be we can do it too.' Twenty-five men's SHGs have been formed so far. They have not only begun to save money, but have bought seeds and fertilizers as a group, thus either getting better deals or bulk discounts. Something else has happened. 'Yes,' one man said sheepishly, speaking for himself and his brothers, 'We have stopped beating our wives.'

WOMEN'S VOICES

These are translations of what the women of Okhamandal said. In each transcript, there is a new voice emerging. Each voice speaks, simply yet powerfully, of women's empowerment.

> Ever since I've joined the mandal and become financially independent things have changed. Earlier, because of lack of money my husband and I used to fight with each other. But today I too contribute to the running of the house and there is peace at home. We both go our ways to work ... where is the time to fight? People's way of thinking has changed now. Families are getting smaller, these days women have either two or three children, not more. We now understand the benefits of having a small family.
>
> I have started going to the beauty parlour as I have the freedom to spend money on myself. My husband encourages me in this. He tells me I should look smart. I've seen women from other villages looking good because of going to the parlour, so I thought why not me? I have the money now. We women today

spend money on gold too. See, we wear a lot of gold. Other things we spend cash on are clothes, primus stoves, and other things for the house.

Rajuben, 28

I've been with a mandal for one-and-a-half years. I come from a well-off family. I am the only daughter of my parents. My husband is an accountant in the cooperative store. I didn't join the mandal for financial reasons. I was always fascinated with the way they changed lives. When I used to attend the meetings of the women mandal earlier, my neighbours used to ask me how I could go for meetings without an invitation, but I never bothered. Because of that I learnt doing zari work. I have taught sewing to women from very poor families. I have joined the mandal to help women and do something for the community.

Jiviben, 45

I've been with Roshni Mahila Mandal for two years. I am a widow and have a daughter. I was working as a helper in a nursery but the money I got from there wasn't enough to run the house. So I was asked to form a mandal. I managed to convince my own relatives to join me, they all wanted to help me so they agreed. But I was told that eight women aren't enough. It was then that I approached women of other communities to join in and they did. There are Rabaris, Rajputs and others in the group. Earlier Rabaris wouldn't come to our house and we could never go to their homes or even touch them, but because of the mandals all that has changed and there is a lot of mingling between us. We are all friends too.

My biggest happiness today is that because of my financial independence I can buy my daughter whatever she asks for. We are really happy today. In addition to coming to the community centre to do the finishing work of garments I make washing powder. I learnt this skill from social workers. I want to save money to have a good bank balance to get my daughter married but at the same time I tell her that she should study and become financially independent. I studied only upto the seventh standard. I want her to work in an office, sit on a chair.

I went to Ahmedabad the other day and I loved going there. My dream is to have enough money to travel all over the country. But on second thoughts I would like to go through Tata on work, then I won't have to spend my own money.

Khatija, 35

I've been with the Vankol Mandal for five years. I got to know about mandals and their activities through Nathiben. We started with twenty women, then it was down to ten for a while but today it has thirty members. The reason it came to ten was that women didn't trust women. They thought their money was not safe with us. But today after seeing the benefits they are more confident. There is a lot of bonding amongst us, now. We share our lives with each other.

We earn from Rs 500 to Rs 1500, but there are no limits. One can earn as much as one works. Sewing is a skill we learn at a very young age and today it is earning us our livelihood. Most of us aren't literate, but we've learnt to sign our names and yes we are good at counting money. No one can cheat us on that.

Lakhiben, 30

Our mandal is eight years old. Before coming to the mandal I worked on the farm. I lived like an uncivilized person, the only place I went to if I stepped out of my house was the jungles! Today I am leaving for Mumbai for the exhibition of our works. It all began when the Tata people came to our village to start the water programme for the farmers, some of the women suggested that I start a mandal.

Rehanaben from Tata taught me to sign my name. I managed to get ten women to join. We started off with Rs 10 per member, then we felt ten was too less and we increased the contribution to twenty. After a year of saving, Tata Chemicals gave me a loan of Rs 7000 and we also got a loan from Bank of Baroda. We bought a thresher with the money. It was bought to generate income. We gave it out on rent to the villages and soon we not only recovered our money but also paid off the loan much before the due date.

There are thirty-two women in my mandal. Each member pays Rs 20 every month. We run a nursery where neem saplings are sold. We buy seeds from the wholesale market and sell to the farmers here at retail rates. We make a profit of Rs 10 per kilo. The Tata people linked us to the distributors from the wholesale market.

The mandal today has Rs 150,000 in its kitty and can easily give out loans. My mandal helps women in their children's education, for medical purposes, to build/buy a house, buy seeds. I earn about Rs 1500-2000 a month by sewing. The women of our mandal today send their children to schools. We had never seen a bank before, today we confidently go there. My children have studied upto class twelfth. My oldest son runs a shop, the second one is doing teacher's training and the youngest is in degree college. My daughters have also studied upto ninth std. One helps me in the sewing and tailoring work, while the other takes care of the house. My elder daughter also earns Rs 1500.

I've been to Jamnagar, Anand and Ahmedabad for mandal work. The first place I visited was Ahmedabad. I remember when I was to go there, the village menfolk told my husband that I will never come back. Forty of us were going. Mr Padi who was taking us was angry with these remarks, he told them that he would pay all of them Rs 90,000 if their wives didn't come back. That is the usual bride price. He said, "If the women don't come back I will give you Rs 90,000 and you can get new wives!"

I was nervous about travelling anywhere then. Now I am not nervous of going even to Mumbai. I am very confident today. My way of speaking has changed. Somebody told me that since I am going to Mumbai for the exhibition, what if a film star like Madhuri Dixit came to the exhibition and liked something. I said well I would tell her to take what she liked but make sure that she paid for it!

I am confident of talking to strangers. Earlier when people like you came to talk to us, we would wish they went away soon. We wouldn't respond well to discourage them. But my husband changed all that. He organized a meeting for us with Pragnaben, one of the social workers. He made us understand that it was all for our good. Today I have the confidence to ask you to share my lunch with me! Another thing that has changed for us is that the men take us into confidence while discussing village matters. Our opinions are considered and we too are not shy of speaking to village elders, in whose presence at one time we couldn't stand even for a minute.

Our food habits have also changed. Earlier we ate only jowar and bajri rotlas with butter. Today we eat dhal, rice and vegetables. We are aware of the nutritive value of vegetables. Health awareness has increased and we keep our homes neat and clean. Earlier when we would fall sick we would go to mataji, but once my son had a bad case of tetanus and inspite of being under mataji's treatment for five days when his condition didn't improve we took him to the hospital. From then on we go only to the doctor at the hospital. In the old days when we got a headache we would smear mud on our forehead. We don't do that now.

As I mentioned earlier, the bride price in our village was Rs 90,000. But we now have a standard rate of eight tolas of gold and two sets of clothes. All the people of our community have to follow this. Also for a marriage engagement in the old days two truckloads of people would come as guests, today it is only five people. All these changes have been first thought of by us women and the men have followed.

What I want to happen in the future is to connect all the mandals. I want more schools and hospitals to come up here in my village itself so that our children don't have to leave their homes for studies. I know all this will be possible. God will help us when we help ourselves.

Nathiben, 40. Mordav Mahila Mandal

GIVING LIFE TO LIFE

Mithapur is near Dwarka. Dwarka is where the Blue God, Krishna, set up his capital after leaving Mathura. How appropriate, then, that there is a story of Krishna and salt.

Krishna had two wives, Satyabhama and Rukmini. Satyabhama was beautiful and vain, but she was insecure too. Did her husband love Rukmini more than her? One day she put the question to him, point blank. 'Whom do I love more?' Krishna answered, 'Let me put it this way: You are like sugar, Rukmini is like salt.' This pleased Satyabhama no end. 'After all', she said to anyone who would listen, 'everyone loves sweets. Who likes salt?'

One day when there were guests for dinner, Krishna got the cook to bring in only sweet dishes. The guests ate a few, then looked at each other. Was this a meal? They grumbled under their breath. Krishna then signalled the cook and in came one dish which had salt in it, then another, then yet another. The guests fell on the food greedily, licking their lips, complimenting the chef.

Krishna smiled at Rukmini. Then he smiled at Satyabhama. Satyabhama didn't smile back: She now understood what Krishna had really meant.

Harivamsa, the epilogue to the *Mahabharat*, describes the land near Dwarka 'overflowing with vegetation, a place full of deer and even elephants.' The elephants and deer have gone, along with the lush vegetation of those days, but Okhamandal has life of another kind in plentiful supply.

Located at the tip of the Saurashtra peninsula, Okhamandal is surrounded by water on three sides: there's the Gulf of Kutch to the north and the Arabian Sea to the west and the south. The coastline has bays and beaches, coves and cliffs and many small islands. At one point on the road, you suddenly realize you are at a height. The ground slopes rapidly down to a crescent-shaped beach, near which are dense growths of mangrove. These send down roots from their branches so that a network of many stilt-like roots supports the leafy part of the mangrove. These roots in their hundreds catch silt, which then accumulates in the water. This acts like a kind of bund, slowing down the incoming tide, thus stopping erosion. Mangrove seeds often germinate while the fruit is still on the tree. A seed will send down a root 30 cm (one foot) long. This will strike earth and a new tree begins to grow. The roots serve yet another purpose: they become a breeding place for fish and marine life.

There's more life on the way to Dab Dabha Island, a short boat ride away. Dolphin and marine turtle sightings are not uncommon, though the beaches of Shivrajpur and Khanjani are supposed to be the best places to observe the Olive Ridley and green-sea turtles.

One of the Krishna stories describes a forest at Okhamandal being the home of Som *vel*, the Som creeper from which the gods made Somras, the elixir of youth. No one's found Somras yet, or if they have, they aren't talking, but there are any number of medicinal herbs and plants, some of them rare species which have been located there. The forest is said to have quite a population of animals too: porcupine, wild boar, *neelgai*, jackals and hyenas.

At the Boria and Gugar reefs there's marine fauna and flora, sea anemones, corals, lilies and mollusces. In the nearby islands, there are reported sightings of the Dugongs (sea cows).

But the most amazing sights are provided by the migratory birds who come in profusion to the Charakla Salt Works, the series of salt pans which lead up to the saturated brine that goes into Tata Salt. The surprise visitors were a flock of Caspian terns, which in 2000 decided to set up a colony here and made it into their breeding place. One good tern, as the old joke goes, deserves another.

Tern is a family of sea birds related to gulls. They have long, pointed bills and webbed feet. Their pointed wings are made for real flying: terns fly both fast and long. They are often called sea swallows because of their swift and graceful flight. When they want to feed, they do a quick dive into the water, beak down, and emerge with the catch of a small fish.

The Caspian tern has been described as a handsome bird. It is the largest of the tern family, some 53 cm (21 in) long, with a shining black crest and pearl-grey back and wings. It lays its eggs in nests of seaweed, or sometimes even on bare rock. The parents take turns (is that how the birds got their name?), sitting on them till they hatch. Their sudden choice of this area for breeding took everyone by surprise: *The Field Guide to Waterbirds of Asia*, for example, says that they breed in Central Asia and move to South Asia during their non-breeding period. Now, every winter, as many as five hundred Caspian terns can be seen at their new adopted home at the saltworks.

That's small compared to some of the other kinds of birds which come to this part of the country. As many as 121 species of birds have been spotted at Charakla. (The full list is in an appendix.) The more prominent of them, and their average strength each year is

Greater Flamingos	:	5,000 to 12,000
Lesser Flamingos	:	2,000 to 4,000 plus
Painted Storks	:	150 to 500 plus

Rosy Pelicans	:	200 to 800
Dalmation Pelicans	:	50 to 200
Shoveller Ducks	:	50 to 100
Blackneck Grebes	:	150 to 2,000
Caspian Terns	:	300 to 500
Reef Heron s	:	50 to 60
Large Egrets	:	60 to 100

The Charakla Salt works is located in the Little Rann of Okhamandal. It is a vast area, spreading over nearly 30,000 acres, stretching from the Gulf of Kutch to the Arabian Sea coast. It becomes an important stopover point for birds from two directions. There's the extensive migration from the Central Asain region which takes place during autumn every year. The birds fly down the Indus plain and split up over the Kutch area in two directions: either to fly south to peninsular India and Sri Lanka or to the west along the Mekran coast, down the Arabian coast to East Africa. A similar 'flyway' exists for birds from Easter Europe: they come over the plains of the River Tigris, the Eupharates and the Persian Gulf.

But geographical location isn't the only reason the birds come here. To start with, the extensive mangrove vegetation provides ideal shelter. Another attraction is the diverse physical features of Charakla – the shallow sea water ponds; the tidal mudflats; the mangrove belt on the Gulf of Kutch coast; the scrub vegetation on the flanks; and the fallow lands – these provide diverse habitats that support an equally diverse birdlife.

Another important reason is the plentiful food supply: they get feed in the tidal mud-flats along the Gulf of Kutch shoreline, they also get a rich harvest because of the regular pumping in of seawater into the salt pans, which brings in fresh quantities of small fish, fish eggs, small marine creatures, brine shrimps, mollusks, algae and even fresh prawns. Since 500,000 kilolitres of sea water is pumped in every day, the food supply never dries up.

One more reason for the birds' preference for Charakla is that this area is a safe haven for them. Tata Chemicals has taken on itself the job of ensuring there is no poaching. The birds obviously feel safe; they do not react to the sound of cars driving along the salt pans, or the click of the camera shutter.

The most striking of all the birds at Charakla are the flamingos. They stand out, literally, because of their height, which can go up to 150 cm (5 feet). And that's without stretching their very long necks or their long (and curved) beaks. Then there are their legs, long and stilt-like, which wouldn't be out of place on a fashion ramp. As if this weren't enough, the colour of their feathers, a pale pink (some flamingos, particularly of the Caribbean area have feathers that are coral red).

Flamingos flock together, and it is an unforgettable sight when a whole colony of them come flying in, a large pink swarm descending from the air and fanning out into the water, where they busily keep dipping their bills into water, looking for shellfish and algae.

Hair-like combs on the edges of their bills help to strain out mud and sand from the food the flamingo picks out from the water.

Flamingos mate only once a year and lay only a single egg in a nest made from mud. Both parents take turns to sit on it, which takes a whole month to hatch. A young flamingo, surprisingly, doesn't look like a flamingo at all but like a baby duck with a short beak and no pink feathers. It ventures out of the nest after five days; after fourteen, it is ready to socialize with other youngsters and find its own food. It will then lead its own life for the next fifteen to twenty years.

Both the Greater Flamingo and the Lesser Flamingo come to the salt works. The latter has been put on the Endangered Species list internationally. Hopefully, the safe haven of Charakla will get it off the list.

There are other endangered species which come here. The Dalmation Pelican and the Black-necked Stork have been classified as 'Globally Threatened Species' by international experts.

The Black-necked Stork had become an extremely rare sight in Saurashtra, but since the mid-1990s, it seems to be making a come-back, a sign that conservation efforts are beginning to produce results. It helps that in many communities, this species is supposed to bring good luck. And, of course, fly in with the babies. That particular story probably grew out of the fact that storks are known to take loving care of their young. And also that they look strong enough for the job, with their one metre (three feet) height, long legs, strong wings and pointed beak. Storks, incidentally, are known to return to the same nest year after year, and with the same mate.

If there were another candidate for carrying babies, it would have been the pelican. It is formidably large and its long, straight bill, with the familiar pouch underneath, could probably carry twins in one go. As it happens, the pouches are to keep fish and the pelican needs a lot of fish (upto 3 kg a day) because of its own body weight. It is approximately 1.5 metres (five feet) long; its wing spread can go up to a massive three metres (ten feet).

Pelicans fish in groups. They swim along the top of the water in a line, shepherding the fish ahead of them. After chasing the fish into shallow waters, they scoop up the catch with their pouches. An even better sight is to watch pelicans fly in a V-formation. They soar upwards on a thermal, a rising current of hot air. The leader, the one at the junction of the V, creates 'lift' for the other birds. The leadership position is rotated in mid-flight, a democratic way of sharing the burden of flight.

The black-necked grebe is the biggest success story of Charakla. A few were seen through the years, but their number was so limited that these were obvious vagrants. But since the 1990s, there has been a virtual explosion of them so that their number, ten years later, is in the thousands. In 2004, for example, well over two thousand of these birds had made Charakla their winter home.

Grebes (for some reason, pronounced greebs) are diving birds, diving with such seeming recklessness that they are called Hell Divers. But they aren't reckless; they are just made for diving. Their bodies are flat, and covered with waterproof feathers. Their wings are small and they have a short tail. Their legs are far back on their bodies; they also do not have webbed feet. The net result is that on land, they have very poor balance for both standing and walking. They also have a problem taking-off for flight. But once aloft, they can fly really long distances, or dive and swim effortlessly.

Grebes usually nest in shallow marshes or ponds, which must attract them to salt pans, making their nests of water-grasses and plants. Their young can swim as soon as they are born, but when they need a rest, they sit on their parents' back or nestle under their wings. The parents often dive while they are carrying their young, the only bird to do so. In the West, grebes were killed for their feathers which were used in ladies' hats. Ladies no longer wear hats; in any case, the birds are now protected by law.

More often than not, legislation isn't enough. A wonderful example of how to go beyond laws and thus make them work, is provided by the strange tale of the Great Whale Shark and the coast of Gujarat.

We know of the great whale and we know of the great white shark, one a gentle giant, the other, the great predator. There is something in-between, the great whale shark. It belongs to the shark family and it is larger than the great white; in fact, it is the largest fish in the world (The whale, which is even larger, is not a fish but a mammal). Unlike its violent cousin, the whale shark is almost vegetarian, eating only plankton plus a few hundred small fish which get sucked in, almost accidentally, when it gulps down water.

For hundreds of years, whale sharks have been coming to the Gujarat coast from Australian waters. They come to the waters near Porbander between March and May which is their spawning period.

Its very size and its gentle nature, however, has made the whale shark a prey to two-legged predators. Each fish is worth Rs 100,000 for its meat, liver and other body parts; a sum large enough to tempt so many fishermen that they killed as many as 1200 whale sharks every year. They were killed cruelly too: a fish was 'false-hooked' (i.e., hooks were thrown randomly at any part of the body), barrels were tied to the ropes to keep the injured fish from sinking and it was then towed to shore where it was carved up.

The Indian government made the whale shark a protected species under the Wildlife Protection Act in 2001, banning the catching of the fish and any trade in its meat. But the Gujarat coastline is 1600 km long and policing such a length of coastline effectively was difficult even though the Indian Navy and the Coast Guard joined forces.

A partnership involving four bodies decided to do something beyond this: the Wildlife Trust of India and the International Fund for Animal Welfare tied up with two corporates, Tata Chemicals and Gujarat Heavy Chemicals. The corporates went beyond providing financial assistance; the employees of Tata Chemicals, for example, were encouraged to become volunteers in the programme, and this way a core team of highly committed individuals was formed.

A 'Save the Whale Shark' campaign was launched. This was given a huge boost when the extremely popular spiritual leader of Gujarat, Morari Bapu agreed to incorporate the message in his discourses. He not only reminds his listeners about the Indian tradition of hospitality but also compares the whale shark to a daughter who comes to her parents' home to give birth.

The campaign's message about the cruel fate of the whale shark has been taken into schools, into adult institutions and into streets through street plays and exhibitions. The most effective part of the campaign, however, is the life-size twelve metre (40 ft) inflatable model of the fish, which is carried on a camel cart and is welcomed by locals as they would a deity, with garlands and vermilion. As a result, the whale shark, once contemptuously referred to as *dhol* (after the barrels used for fishing), is now called *vahali*, meaning *pyari* or dear one.

Proof that the campaign has begun to work came in October 2004. A full-grown whale shark got entangled in the fishing nets of a boat off the coast of Dwarka. The owner ordered the knives out. Not to kill the fish, but to cut the nets. That was a double loss in terms of money for the owner. 'But who thinks of money when you want to save a dear one?' the man said.

So here we are then, near Dwarka, Krishna's place: near Porbander, where Gandhi was born: both not far from Dandi where Gandhi's march ended: nor too far from Sabarmati where the ashram is located. Salt has been the common link. Salt, which is in our blood, salt which gives life to us, and which gives life to life. The romance of salt began a long time ago. It is a romance which will never end.

SELECT BIBLIOGRAPHY

Part 1

REFERENCES

1. Joan V. Bondurant, *Conquest of Violence: The Gandhian Philosophy of Conflict* (Bombay, 1959)
2. M.K.Gandhi, *Non-Violence in Peace and War* (Ahmedabad, 1942)
3. H.J.N. Horsburgh, *Non-Violence and Aggression: A Study of Gandhi's Moral Equivalent of War* (London, 1968)
4. E. Stanley Jones, *Mahatma Gandhi: An Interpretation* (London, 1948)
5. Rajendra Prasad, *At the Feet of Mahatma Gandhi* (Bombay, 1961)
6. The Collected Works of Mahatma Gandhi
7. Jawaharlal Nehru, *A Bunch of Old Letters*
8. Nirmal Kumar Bose, *A Study of Satyagraha* (University of Poona, 1968)
9. Louis Fischer, *The Life of Mahatma Gandhi,* (London, 1962)
10. *Young India,* Navjivan Prakash Mandir (Various Issue of 1930)

11. B R Nanda, *Mahatma Gandhi, A Biography* (Oxford University Press, 1958)
12. C F Andrews, *Mahatma Gandhi's Ideas* (Allen & Unwin, London, 1929)
13. Mahadev Desai, *The Diaries* (Navjivan Publishing House, 1953)
14. Kaka Kalekar, *Stray Glimpses of Bapu* (Navjivan Publishing House, 1950)
15. Jawaharlal Nehru, *Mahatma Gandhi* (Signet Press, Calcutta, 1949)
16. Romain Rolland, *Mahatma Gandhi* (Allen & Unwin, London, 1924)
17. D. G. Tendulkar, *Mahatma* (Times of India Press, Bombay)
18. Jyotsna Tewari, *Sabarmati to Dandi: Gandhi's Non-violent March and the Raj* (Raj Publications, Delhi, 1995)
19. Thomas Weber, *On the Salt March* (Harper Collins, New Delhi, 1997)
20. Geoffrey Ash, *Gandhi, A Study in Revolution*

Part 2

1. Kapilram H Vakil, *Salt, Its sources and supplies in India*, The Commercial Printing Press, Bombay 1945
2. Evan Marlett Body, *History of Salt*
3. Dorothy Telfer, *About Salt*, 1967
4. Peire Laszlo, *Salt, Grain of Life* (Columbia Univ. Press)
5. Kapilram H Vakil, *Salt Supplies of Bengal, Bihar, Orissa & Assam* (Indian Salt Associate, 1938)
6. Kapilram H Vakil, *Indian Salt Industry* (Internal Paper, Oct 1936)
7. Roy Moxham, *The Great Hedge of India* (Harper Collins Publishers, New Delhi, 2001)
8. Adam Roberts, *Salt* (Victor Gollancz, 2000)
9. Pablo Neruda, *Elementary Odes*

10. Charles Dickens, *To be Taken with a Grain of Salt* (Strand Christmas Number, 1865)
11. Mark Kurlansky, *Salt, A World History* (Penguin, 2002)
12. Robert Froman, *The Science of Salt*. Mckay, 1967
13. Robert Kraske, *Crystals of Life: The Story of Salt*. Doubleday 1968
14. Ralph Minear, *The Joy of Living Salt Free*, Macmillan, 1984
15. Robert Multhauf, *Neptune's Gift, A History of Common Salt*. John Hopkins, 1978
16. Kapilram H Vakil, *The Okha Salt Works* (Internal Paper, Dec 1936)
17. Ernest Jones, *Essays in Applied Psychoanalysis* (Hogarth Press, 1951)
18. James Frazer, *The Golden Bough: The Roots of Religion and Folklore* (Macmillan, London, 1890)
19. Harry Middleton Hyatt, *Hoodoo, Conjuration, Witchcraft*
20. Joseph Campbell, *The Masks of God* (Harper, 1962)